Ashis K. Mukherjee

The 'Big Four' Snakes of India

Venom Composition, Pharmacological Properties and Treatment of Envenomation

 Springer

Ashis K. Mukherjee
Life Sciences
Institute of Advanced Study in Science
and Technology
Guwahati, Assam, India

ISBN 978-981-16-2898-6   ISBN 978-981-16-2896-2   (eBook)
https://doi.org/10.1007/978-981-16-2896-2

This Springer imprint is published by the registered company Springer Nature Singapore Pte Ltd.
The registered company address is: 152 Beach Road, #21-01/04 Gateway East, Singapore 189721,
Singapore

Preface

Snakes, forming an important component of biota, are intrinsically fascinating and enchanting reptiles. The word "snake" probably evokes a wider range of feeling than any other living creature, both in the sense of admiration and respect and being despised as a symbol of evil. Snakebite itself is an intimidating disaster. A minute amount of toxic material of venomous snakes produced in their specialized salivary glands, known as venom, when injected into the prey or victims, causes intense pain and may lead to death within a very short span of time. This leads to distress, agony, and pathos in the mind of common people, as a result of which snakes and their venoms become shrouded with myths, mysteries, and superstitions. Besides, snake-bite is a serious problem for the rural tropics, most particularly in the Southeast Asian countries. In India alone, more than 100,000 cases of envenomation occur per year (unreported cases may be more than this), which ultimately results in substantial deaths and/or morbidity. Unfortunately, less attention has been paid to provide affordable and effective treatment against snakebite. Due to lack of our knowledge on geographical and species-specific variation in snake venoms' composition and mode of action, lack of good understanding of clinical features of envenoming, and poor efficacy of commercial antivenom produced by Indian manufacturers, snake-bite management remains highly unsatisfactory in the Indian subcontinent. There-fore, the World Health Organization (WHO) has declared snakebite as one of the neglected tropical diseases.

From the last few decades, people have been gradually realizing that the facts about snakes are much more interesting than many popular beliefs, embellishment, and misconceptions. Apart from understanding the molecular mechanism of action of snake venom components and improvement of treatment of snakebite patients, it is equally essential for us to appreciate the scientific contributions of snakes in our ecosystem as well as therapeutic and diagnostic applications of snake venom proteins. Therefore, during the recent years, the subject of snake venom has been of scientific research interest from the perspective of biochemistry, toxinology, pathophysiology, pharmacology, immunology, and biomedical research.

Such a high rate of mortality of snakebite in India as well as the remarkable medical and diagnostic applications of snake venom toxins have prompted me to pursue research activity in this exigent field. The "Big Four" venomous snakes—Indian cobra, Indian common krait, Indian Russell's viper, and Indian saw-scaled viper—have received enormous medical importance in the Indian sub-continent because they are responsible for the majority of snakebite deaths in these regions. This book presents a comprehensive study of venoms of the "Big Four" venomous snakes of India, evolution and variation in venom composition, biochemical and pharmacological properties of venom, biomedical application of snake venoms, and treatment of snakebite patients. I am sure that the book will serve as a standard reference for researchers and medical students as well as for those interested in snake venom. This book will cater to the quest of the readers, and they will be eager to learn more about the "Big Four" venomous snakes and biomedical application in snakebite patients.

It is a great pleasure for me to convey my words of appreciations to pioneers in the field of snake venom research for their useful suggestions, constructive criticism, and kind cooperation in editing the book chapters. I thank my wife Abira and son Anandan for their patience and encouragement and for understanding my idiosyncrasies, and my father Prof. Subhas C. Mukherjee and mother Mrs. (Dr.) Sandhya Mukherjee for their constant inspiration. I duly acknowledge the editing of book chapters by my brother Dr. Soumen Mukherjee, India, and Dr. Glen Wheeler, Canada.

Guwahati, India   Ashis K. Mukherjee

Contents

1 Introduction.. 1
   1.1 A Glimpse of the Venomous Snakes of India.................. 2
   1.2 The Concept of the "Big Four" and Non-"Big Four" Medically
        Important Venomous Snakes of India........................ 4
   1.3 Medical Aspects of Snakebite: The Snakebite Problem......... 6
        1.3.1 Snakebite in Developed Countries................... 8
        1.3.2 Epidemiology of Snakebite in Asia.................. 9
        1.3.3 Epidemiology of Snakebite in India................. 10
   1.4 Key Issues Pertaining to Snakebite in India................ 15
   References.. 16

**2 Evolution of Snakes and Systematics of the "Big Four" Venomous
   Snakes of India**... 21
   2.1 Evolution of Snakes....................................... 22
   2.2 Studies of the Genomics, Phenomics, and Fossil Record Show
        the Origin and Evolution of Snakes....................... 24
   2.3 Studies on the Genomic Regression of Claw Keratin, Taste
        Receptors, and Light-Associated Genes and the Evolutionary
        Origin of Snakes... 24
   2.4 Skull Evolution and the Ecological Origin of Snakes........ 26
   2.5 Systematics of the "Big Four" Venomous Snakes of India..... 26
   2.6 The "Big Four" Venomous Snakes of India Represent the
        Advanced Group of Snakes................................. 28
        2.6.1 Family Elapidae................................... 29
        2.6.2 Family Viperidae.................................. 30
   References.. 32

**3 Snake Venom: Composition, Function, and Biomedical
   Applications**.. 35
   3.1 The Venom Gland and Venom Delivery Apparatus in the
        Viperidae and Elapidae Families of Snakes................ 36
   3.2 Comprehensive Review of the Venom Composition of the
        "Big Four" Venomous Snakes of India...................... 39
        3.2.1 Enzymatic Toxins of the "Big Four" Snake Venoms.... 40

        3.2.2   Nonenzymatic Toxins in the "Big Four" Snake
                Venoms   41
        3.2.3   Nonprotein Components of Snake Venom   45
   3.3 Variation in Snake Venom Composition and Its Impact on the
        Pathogenesis of Snakebite and Antivenom Treatment   49
   3.4 Evolution of Genes for the Toxins in Snake Venom   50
        3.4.1   Toxicofera Hypothesis   50
        3.4.2   Independent Origin Hypothesis   50
   3.5 Mechanism of the Evolution and Diversification of Venom
        Proteins   51
        3.5.1   Accelerated Evolution of Venom Protein Genes   51
        3.5.2   Selection Pressure for Rapid Adaptive Evolution   52
        3.5.3   Diet and Snake Venom Evolution   52
   3.6 Biological Functions of Venom   52
        3.6.1   Prey-Specific Venom Toxicity   53
        3.6.2   Immobilization and Predigestion of Prey   53
        3.6.3   Prey Re-localization   53
   3.7 Indian Snake Venom Proteins: A Treasure House of Drug
        Prototypes and Diagnostic Tool   54
   References   58

4 Indian Spectacled Cobra (*Naja naja*)   69
   4.1 Taxonomic Classification of the Indian Spectacled Cobra
        (*Naja naja*)   70
   4.2 Characteristic Features of the Indian Spectacled Cobra   70
   4.3 Geographical Distribution and Reproduction of the Indian
        Spectacled Cobra   73
   4.4 Biochemical Composition of the Indian Spectacled Cobra
        Venom   74
   4.5 Biochemical and Proteomic Analyses to Demonstrate the
        Geographical Differences in Venom Composition of Indian
        Spectacled Cobra (*N. naja*) Venom   82
   4.6 Genomic and Transcriptomic Analyses of Indian Spectacled
        Cobra Venom Toxins   84
   4.7 Species-Specific Differences in the Venom Composition Between
        N. naja and *N. kaouthia* from the Same Geographical Location of
        the Country   86
   4.8 Pharmacology, Pathophysiology, and Clinical Features of Indian
        Spectacled Cobra Envenomation   88
   References   90

5 Indian Common Krait (*Bungarus caeruleus*)   95
   5.1 Taxonomic Classification of the Indian Common Krait
        (*Bungarus caeruleus*)   96
   5.2 Characteristic Features of the Indian Common Krait   96

5.3 Geographical Distribution, Habitat, Behavior, and Reproduction
of the Indian Common Krait............................. 97
5.4 Venom Composition of the Indian Common Krait............ 98
5.5 Pharmacology, Pathophysiology, and Clinical Features of the
Indian Common Krait Envenomation...................... 101
References.. 102

6 Indian Russell's Viper (*Daboia russelii*)...................... 105
6.1 Taxonomic Classification of Indian Russell's Viper
(*Daboia russelii*).................................... 106
6.2 Characteristic Features of the Indian Russell's Viper........... 106
6.3 Geographical Distribution, Habitat, and Reproduction of Indian
Russell's Viper....................................... 108
6.4 Composition of Indian Russell's Viper Venom............... 109
6.5 Pharmacology, Pathophysiology, and Clinical Features of
Envenomation by Indian Russell's Viper................... 123
References.. 128

7 Indian Saw-Scaled Viper (*Echis carinatus carinatus*)............. 135
7.1 Taxonomic Classification of the Indian Saw-Scaled Viper
(*Echis carinatus carinatus*)............................ 136
7.2 Characteristic Features of the Indian Saw-Scaled Viper......... 136
7.3 Geographic Distribution, Habitat, Behavior, and Reproduction
of the Indian Saw-Scaled Viper.......................... 137
7.4 Composition of the Indian Saw-Scaled Viper Venom........... 138
7.5 Pharmacology, Pathophysiology, and Clinical Features of
Envenomation by the Indian Saw-Scaled Viper.............. 141
References.. 143

8 Prevention and Treatment of the "Big Four" Snakebite in
India... 145
8.1 Prevention of Snakebite: Some Useful Strategies............. 146
8.2 First Aid for Snakebite.................................. 147
8.2.1 First Aid for Snakebite: World Health Organization-
Recommended Guidelines......................... 147
8.3 Antivenom Production in India........................... 148
8.3.1 Monovalent vs. Polyvalent Antivenom................ 148
8.3.2 Production of F(ab')2 PAV in India.................. 149
8.3.3 Quality Control of Commercial Antivenom: World
Health Organization Guidelines.................... 150
8.4 Diagnosis and Clinical Treatment of Snakebite.............. 152
8.5 Management of Adverse Effects of Antivenom.............. 155
8.5.1 Early Adverse Reactions.......................... 155
8.5.2 Endotoxin-Mediated Pyrogenic Reactions............. 155
8.5.3 Late Serum Reactions............................ 155

        8.5.4   Prevention and Treatment of Adverse Serum
                Reactions   156
   8.6   Geographical and Species-Specific Variation in Snake Venom
         Composition and Its Impact on Antivenom Treatment   157
References   159

About the Author

Ashis K. Mukherjee is the Director of the Institute of Advanced Study in Science and Technology, Guwahati, Assam, India, and also a Professor of Molecular Biology and Biotechnology at Tezpur University, Assam, India. He did M.Sc. (Biochemistry) from Banaras Hindu University, Ph.D. in Biochemistry and Pharmacology of Indian Cobra and Russell's Viper Venom from Burdwan Medical College under Burdwan University, and D.Sc. (Biotechnology) from Calcutta University on characterization and biotechnological application of phospholipase A_2 and proteases from Indian cobra and Russell's viper venom. Prof. Mukherjee has more than 25 years of research experience on Indian snake venoms and treatment of venomous snakebites. His current research interest includes biochemical, pharmacological, and proteomic analyses of Indian snake venoms; quality assessment of commercial antivenom; and novel diagnostics and drug discovery from natural resources including snake venom. He has published numerous research papers in peer-reviewed national and international journals and book chapters, guided Ph.D. students, and received several awards and medals for his academic competence and research achievements, the most notable being Visitor's Award for Research in Basic and Applied Sciences from the Honorable President of India in 2018. Prof. Mukherjee is also the task force member of the Department of Biotechnology, Ministry of Science and Technology, Government of India; Indian Council of Medical Research, Government of India; and the World Health Organization (WHO) on prevention and control of snakebite envenoming.

Introduction

Abstract

Snakes are legless, elongated, carnivorous reptiles of the suborder Serpentes' group. More than 3400 living species of snakes are distributed in most parts of the world in a wide variety of habitats. India has a very rich diversity of snake fauna where approximately 60 species of snakes are venomous and among them, the famous "Big Four" venomous snakes, namely Indian spectacled cobra (*Naja naja*), Indian common krait (*Bungarus caeruleus*), Indian Russell's viper (*Daboia russelii*), and Indian saw-scaled viper (*Echis carinatus*), are found almost all over the country, except in few regions. These "Big Four" venomous snakes are accountable for majority of envenoming in the Indian subcontinent and they are considered as class I medically important venomous snakes of the country, the bite of which requires immediate medical treatment. In addition to the "Big Four" several other species of venomous snakes, for example, Indian monocellate cobra, Wall's krait, Sind krait, king cobra, and several species of pit vipers, are also inhabitants of different regions of India but because of less frequency of bite by these snakes they have received poor medical attention. The epidemiological study on the global incidence of snakebite shows that the incidence of snakebite as well as bite deaths in developed countries are much less compared to those in developing countries. South Asia, followed by Southeast Asia, and East sub-Saharan Africa have recorded the highest incidence of snakebite. Notably, India has the highest incidence of snakebite in the world. Poor attention, inappropriate healthcare system, and scarcity of antivenom are the major reasons for high rate of morbidity and/or mortality post-snake envenomation in the developing nations; therefore, snakebite is declared as a neglected tropical disease by the World Health Organization. Because of the lack of a proper coordinated survey on snakebite, poor maintenance of hospital records, and dearth of awareness among people, it is very difficult to envisage the type(s) of snakebite in a particular region of the country. In conclusion, it may be suggested that a combined effort from clinicians, toxinologists, antivenom manufacturing companies, and health authorities, along with awareness among mass, can certainly eradicate the snakebite problem.

Keywords

Big Four venomous snakes · *Bungarus* · *Daboia* · *Echis* · Epidemiology of snakebite · Indian cobra · Indian krait · Indian Russell's viper · Indian saw-scaled viper · Medically important snakes of India · *Naja* · Non-Big Four venomous snakes · Snakebite

1.1   A Glimpse of the Venomous Snakes of India

The snake is undoubtedly one of the most mysterious, enchanting but misunderstood creatures of the world! The word "snake" invokes in the mind of the people a curious mixture of scare and admiration. Fear is evoked because the snake venom is deadly; the venom, when injected, causes a variety of pathophysiological dysfunctions in the body of the victim, which most often leads to death or morbidity. *Ophidiophobia* (fear of snakes) is a response that we have possibly inherited from our ancestors. Simultaneously, however, snakes have been worshipped as a deity and have been shown great veneration by many communities in different parts of the world, including India (Fig. 1.1).

The Sanskrit name of snake is "Bhujanga" or "Sarpa." From ancient times, snakes have been envisaged to be an integral part of Indian cultural heritage. The cobras are garlands of Lord Shiva (the Hindu deity of creative power) (Fig. 1.2a) whereas Lord Vishnu (the Hindu deity of restoration) sleeps on a thousand-headed snake known as "Shesha (means one which remains) Nag" (Fig. 1.2b). The mythological text of India describes that "Shesha Nag" holds the universe. Again, "Devi Manasa" (the Hindu goddess of snakes) is worshiped mostly in West Bengal, Tripura, and Andaman and Nicobar Islands and some other parts of Northeast India, mainly to

Fig. 1.1 Snake as a symbol of deity in temples of India (photography by the author)

a   b

Fig. 1.2 Indian mythological pictures showing (**a**) snake as the garland of Lord Shiva and (**b**) Lord Vishnu sleeping on a thousand-headed snake (sketch by Mr. Anandan Mukherjee)

get rid of snakes and for curing of snakebite. The "Nag Panchami" is celebrated on the fifth day of bright half of lunar month of "Shravana" (generally in the month of August) to worship and pay homage to the snakes throughout the country; nevertheless, it is more popular in northern India. Nonetheless, Indian culture does not allow indiscriminate killing of snakes and it is often considered as a debauchery.

As a subject, perhaps snakes and snake venom has been far more fascinating than many other subjects for the scientific community. At the same time, however, many of us are not yet entirely free from prejudice about snakes. Consequently, we seem to have failed to sufficiently highlight the bright sides of this innocent creature, mainly due to our ignorance over knowledge about snakes. It is important that we understand the molecular mechanism of snake venom components in inducing toxicity, but it is equally important that we should recognize the role played by snake venom proteins in therapeutics. The countless contribution that snakes make to our echo system, and the great help they render to our farmers by being an integral part of the pest management system in feeding rodents, is noteworthy.

Snakes are legless, elongated, carnivorous reptiles of the suborder Serpentes' group. The lack of eyelids and external ears can very well distinguish snakes from legless lizards. Today, there are more than 3400 living species of snakes (serpents) on earth, distributed in most parts of the world and inhabiting fossorial (primarily lives in underground), arboreal (living in trees), terrestrial (on earth), and aquatic (water bodies) environments (Hsiang et al., 2015). Snakes inhabit desserts, tropical rain forests, as well as oceans. India has a very rich diversity of snake fauna, of which only 250 species have so far been identified (Sharma, 1998). Among them, 60 of these species are either venomous or harmful (Sharma, 1998).

The evolution of snakes is disputed and is an enduring mystery. Every venomous organism, such as snakes, is armed with venom gland—the primary function of which is to synthesize and store a myriad of toxins, for performing some important functions for the snake. The detailed description on the composition and evolution of snake venom is given in the next chapters. Snakes use their venom primarily as an offensive weapon to kill, incapacitate, and paralyze their prey, for example, agile rodent and flying bird, before swallowing it. They use their venom as a defensive strategy against predators and rarely against humans. This is the secondary use of snake venom, which includes its use as an aid to the digestion of the food (Weinstein et al., 2010); however, we know very little about the role of natural selection in the evolution of venoms.

In recent years, the subject of snake venom has been ever increasingly important in biochemistry, toxicology, pathophysiology, pharmacology, immunology, and biomedical research. It is noteworthy to mention that the use of snake venom is not limited only to the production of antivenom for saving the lives of snakebite patients albeit snake venom toxins have a huge potential use in the treatment of diseases, some of which has been realized in the invention of novel drugs. The therapeutic applications of components of the "Big Four" snake venoms have been described elsewhere in this book.

1.2   The Concept of the "Big Four" and Non-"Big Four" Medically Important Venomous Snakes of India

The planet is home to more than 3000 species of snakes inhabited across every parts of the world, except Antarctica (Gold et al., 2002; Kasturiratne et al., 2008). However, only about 600 species of snakes of the world are venomous and therefore most of them are nonvenomous (Gold et al., 2002). More than 250 species of snakes are endemic to India; nonetheless only 60 species are venomous (Sharma, 1998; Whitaker et al., 2004; Mohapatra et al., 2011) and among them, the famous "Big Four" venomous snakes (Fig. 1.3), namely Indian spectacled cobra (*Naja naja*), Indian common krait (*Bungarus caeruleus*), Indian Russell's viper (*Daboia russelii*), and Indian saw-scaled viper (*Echis carinatus*), are found all over the country, except in few localities.

Fig. 1.3 Photographs of Big Four venomous snakes of India. (**a**) Indian spectacled cobra *Naja naja* (PC Mr. Romulus Whitaker); (**b**) Indian common krait *Bungarus caeruleus* (PC Mr. Vivek Sharma); (**c**) Indian Russell's viper *Daboia russelii* (PC author); (**d**) Indian saw-scaled viper *Echis carinatus* (PC Mr. Romulus Whitaker; this picture is reproduced from Patra et al., 2017)

The Big Four snakes are accountable for the majority of envenoming that results in morbidity and/or mortality in India and in its neighboring countries (McNamee, 2001; Whitaker, 2006). The Indian cobra (*N. naja*) is uncommon in Northeast India and similarly *E. carinatus* is not found in eastern and north-eastern India. Therefore, bite by these species of snakes is highly unlikely in these regions. However, considering the all-India scenario, the "Big Four" snakes have received a considerable medical attention not only in our country but also in neighboring countries and equine polyvalent antivenom, the only acceptable therapy against snakebite, is available only against the venom of these species of snakes (Mukherjee et al., 2021).

However, the concept of the "Big Four" is highly controversial and criticized by the scientific community and the reason behind is that apart from them, some other species of venomous snakes, for example, Indian monocellate cobra (*N. kaouthia*), Wall's krait (*B. walli*), Sind krait (*B. sindanus*), king cobra (*Ophiophagus hannah*), several species of pit vipers (*Hypnale hypnale, Protobothrops* spp.), and *E. sochureki*, also dwell in different locales of the country and may cause fatalities (Mukherjee & Maity, 2002; Kochar et al., 2007; Joseph et al., 2007; Sharma et al., 2008; Pillai et al., 2012; Warrell et al., 2013; Senji Laxme et al., 2019; Kalita & Mukherjee, 2019). Nevertheless, occurrence of snakebite by these species of snakes, excluding *N. kaouthia* in eastern and north-eastern India, is not reported as compared to bite and death by the "Big Four" venomous snakes (Warrell et al., 2013; Kalita & Mukherjee, 2019). The sea snake (*Hydrophiinae* sp.) and king cobra can also cause lethal envenomation albeit their frequency of contact with human is extremely low; therefore, the number of fatalities from the bite of these snakes is relatively low and negligible. On the contrary, the majority of snakebite patients attending a health center of Himachal Pradesh for snakebite treatment displayed hemotoxic symptom which was correlated with the prevalence of green pit vipers in this region (Gupt et al., 2015). The polyvalent antivenom against the "Big Four" snake venoms did not work well against green pit viper envenomation (Gupt et al., 2015). There was a report showing life-threatening envenoming by the hump-nosed pit viper (*Hypnale hypnale*) in Kerala, south-western India (Joseph et al., 2007). However, a major problem has been encountered that due to lack of snakebite detection kit the hump-nosed pit viper-envenomed patients were also sometimes misidentified by treating physicians as saw-scaled viper (Simpson & Norris, 2007). Similarly, systemic envenoming due to *E. sochureki* is a severe problem in Rajasthan and antidote against this species of snake is not included in Indian commercial polyvalent antivenom (Kochar et al., 2007). Identically, Levantine viper (*Macrovipera lebetina*)-envenomed patients in Jammu and Kashmir show clinical signs of necrosis and hemostatic manifestations and antivenom against this species is also not available in the market (Sharma et al., 2008). Therefore, treatment with Levantine viper antivenom against venomous heterologous species of snakes is debatable and may not provide the best treatment against envenomation.

Therefore, although the decade-long concept of the "Big Four" venomous snakes of medical importance still exists in India, from the aforementioned discussion it seems that this perception may need to be reviewed for the treatment of all types of venomous snakebites across the country. To achieve this much-needed objective, it is extremely essential to have a well-coordinated, national level hospital-based as well as door-to-door snakebite survey initiative to explore all the venomous snakes of India, epidemiology, and frequency of snakebite in each and every remote corner of the country. The results of such survey will assist the health authorities and policy makers of India to make a sound policy which proclaims that besides the "Big Four" venomous snakes of India, the other species of snakes should also be incorporated in the list of medically important snakes to eradicate the snakebite problem forever.

As an alternative concept to the "Big Four," the World Health Organization in 1981 proposed the following for the identification of snakes of medical importance in India (WHO, 1981):

Class I: Snakes commonly cause death or serious disability. Examples include Indian cobra, Indian Russell's viper, and Indian saw-scaled viper.

Class II: Frequency of bites is rare but envenomation by this class of snakes causes serious pharmacological effects leading to local necrosis and/or death. Examples are Indian krait, king cobra, and Levantine viper.

Class III: Bites by this class of snakes may be common; however, they produce least toxic effect in humans and are therefore nonlethal. Example is white-lipped pit viper.

1.3   Medical Aspects of Snakebite: The Snakebite Problem

Estimates show that about five million people around the world are bitten by venomous snakes annually, thus resulting in about 100,000 fatalities (Chippaux, 1998). The latest global epidemiological study on snake envenoming conducted in 227 countries demonstrates that worldwide approximately 421,000 people are bitten by snakes that results in almost 20,000 snakebite deaths annually (Kasturiratne et al., 2008). In many countries of the world, snakebite cases are not systematically reported; therefore, the actual snakebite death may be as high as 94,000 annually (Kasturiratne et al., 2008). Notably, only few countries of the world possess a reliable epidemiological reporting system that can provide precise data on snakebites. Therefore, the magnitude of the snakebite problems has to be assessed through only available scientific reports and literatures which may be more reliable (Chippaux, 1998; Kasturiratne et al., 2008; Gutiérrez et al., 2010). Consequently, due to lack of data, the true global incidences of snake envenoming, death rate, and associated complications such as morbidity are difficult to estimate which is astonishing and alarming (Kasturiratne et al., 2008; Gutiérrez et al., 2010). Recently, by literature analysis on snake envenoming and modeling based on regional estimates of snakebite and deaths, a new method has been developed for an up-to-date estimate of the global burden of snakebite (Kasturiratne et al., 2008; Gutiérrez et al., 2010). On the basis of their findings, the authors have concluded that morbidity and mortality due to snakebite are a serious concern worldwide; however, the highest burden is experienced in South Asia, Southeast Asia, and sub-Saharan Africa (Fig. 1.4).

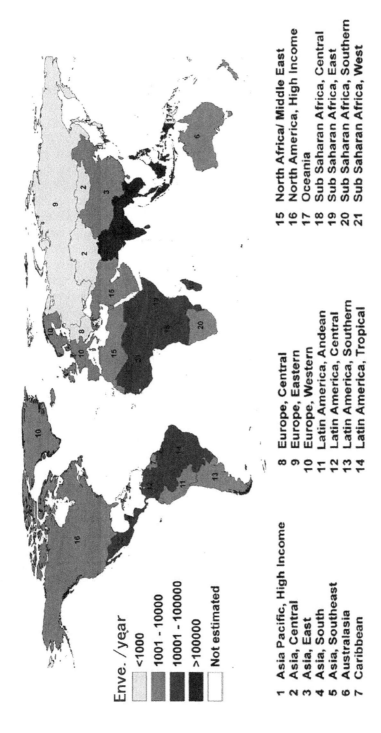

Fig. 1.4 Countries with data on snakebite envenoming (this figure is adopted from the original picture published by Kasturiratne et al. (2008) PLoS Medicine 5, 1591–1604, doi:https://doi.org/10.1371/journal.pmed.0050218.g004)

Snake envenoming is also an occupational health problem that primarily affects the countryside agrarian workforces in developing countries of Oceania, Africa, Asia, and Latin America (WHO, 2007). Conventional snakebite analysis has shown that South Asia (121,000) has the highest incidence of snakebite followed by Southeast Asia (111,000), and East sub-Saharan Africa (43,000) (Kasturiratne et al., 2008; Gutiérrez et al., 2006; Alirol et al., 2010). According to the report presented by Kasturiratne et al. (2008) India has the highest incidence of snakebite in the world where approximately 81,000 envenomings happen annually, which is followed by Sri Lanka (33,000), Vietnam (30,000), Brazil (30,000), Mexico (28,000), Pakistan (20,000), and Nepal (20,000). A recent study has revealed that approximately 314,000 snakebites occur annually in sub-Saharan Africa that results in 7300 deaths and 6000 amputations (Chippaux, 2011). Nevertheless, the problem of snakebite is not appreciated or given prominence in West Africa (Habib et al., 2015). Poor healthcare system coupled with scarcity of antivenom is a major reason for high rate of morbidity and mortality post-snake envenomation in these regions (Warrell et al., 2013). Notwithstanding its corporeal, psychosomatic, and socioeconomic influences, the snakebite problem, its treatment, and research on snake venom have received limited consideration from regional and national health authorities all over the world. Subsequent to the recognition of snakebite as a neglected tropical disease by the World Health Organization (WHO) in 2011, a worldwide awareness and interest have emerged in understanding and addressing this momentous problem. In fact, a major emphasis should be given on the development, standardization, and optimization of methods for reporting the snakebite incidence. Further, a proper record-keeping on morbidity and mortality owing to snake envenomation in healthcare centers, most particularly in the developing countries to confront this disease, would be a most welcome approach (Kasturiratne et al., 2008; Mathers et al., 2007).

1.3.1 Snakebite in Developed Countries

Snakebites, in Europe, are rarely a medical emergency but sometimes may produce severe complications (Chippaux, 2012). In 2012, it was reported that Europe (including European Russia and Turkey) had a total number of 7992 snakebite cases, out of which approximately 15% of the bites were found to be severe (Chippaux, 2012). In Great Britain approximately 200 people are hospitalized every year following snakebite, albeit since 1975, no deaths have been recorded. Conversely, in France, the snakebite incidence is higher than Great Britain (Warrell, 1996). As mentioned by Chippaux et al. (1995), approximately 5 cases per 100,000 residents are reported every year in the Department of Yonne (150 km south of Paris). The rate is similar elsewhere in the country. Very low morbidity due to snakebite is observed in Switzerland, where annually approximately 0.1 case of envenomation occurs per 100,000 populations (Stahel et al., 1985). In the rural areas of Southern Europe, the morbidity rates due to snakebites are higher. In Italy as well as in Spain, the annual occurrence of snakebites can reach 5 per 100,000 people.

Mortality of 0.2 cases per 100,000 people was recorded in Costa Rica during 1990–1993 (Rojas et al., 1997). In France, *Vipera berus* and *Vipera aspis* cause life-threatening envenomation demanding immediate medical treatment (De Haro, 2012).

In Canada and the USA, the annual incidence of snakebite is similar to that observed in Europe. In North America, approximately 45,000 snakebites are reported every year, out of which 15 cause death to the victim (da Silva et al., 2003). In Central and South America, the frequency of snakebite is considerably higher. In Brazil, during 1990–1993, about 20,000 snakebite cases were reported, out of which 90 cases were fatal (da Silva et al., 2003). In North America, approximately 45,000 snakebite cases are reported annually; nevertheless, approximately 10,000 bites are by venomous species. In that, 6500 envenomations require urgent medical treatment, albeit due to advanced medical facility the annual mortality rate of snakebite is meager in the USA (Parrish, 1966; Russell, 1980).

In Australia, the snakebite cases are very few; the annual number of cases that demand medical treatment is not more than 300, which includes 1–4 fatalities (Steward, 2003). Although snakebite frequencies are gradually decreasing in temperate Australia, snakebite-induced morbidity in tropical Australia is considered to be significant (Cheng & Currie, 2004). As is mentioned by Sasa and Vazquez (2003), a total of 5550 snakebites were reported in Costa Rica during 1990–2000. In these snakebites, people living in rural areas, more specifically agricultural workers, were the major victims. Among the venomous snakes, the lancehead (*Bothrops asper*) was found to be mostly responsible for the high snakebite mortality in Cost Rica (Sasa & Vazquez, 2003).

1.3.2   Epidemiology of Snakebite in Asia

Snakebite reports from Asia are higher than those reported from European countries. In Asia, depending on the human activities and snake species involved, an extensive discrepancy in snakebite incidence has been observed. According to the reports of the World Health Organization (WHO, 2011), four million people are bitten by snakes every year in Asia. Half of these bites are by venomous snakes. The annual rate of death from snake bite in India can be estimated at 10,000 (Jena & Sarangi, 1993).

Snakebite is a major problem in the rural tropics such as in India, Bangladesh, Nepal, Bhutan, Pakistan, and Thailand indicating that this is a very serious problem in the Indian subcontinent (WHO, 2011). In Nepal, for example, 3189 cases of snakebite and 144 cases of snakebite death were reported between January and December 2000 from 15 district hospitals of that country (Sharma et al., 2003). The overall death rate of snakebite was 4.5% (Sharma et al., 2003). Between January and December 2000, 4078 cases of snakebite, occurring in different parts of Nepal, were analyzed for clinical and epidemiological features. About 379 of these cases had features of envenoming resulting in the death of 81 people (Sharma et al., 2003). In Taiwan, snakebite record shows that *Bungarus multicinctus*, *Naja atra*,

Trimeresurus mucrosquamatus, Trimeresurus stejnegeri, Deinagkistrodon acutus, and *Daboia russelii siamensis* are accountable for most of the snakebite deaths (Hung, 2004). However, very often snakebite deaths go unreported; consequently hospital records do not give the actual picture of snakebite mortality rate. For example, when the data of snakebites during 1999–2003 as available in the hospitals of the Monaragala District of Sri Lanka was analyzed, it was found that the actual number of snakebite deaths during that period was much higher as compared to the data recorded in the hospitals (Fox et al., 2006).

1.3.3   Epidemiology of Snakebite in India

Swaroop and Grab (1954) initiated a systematic epidemiological study of the snakebite problem in India. They statistically analyzed the countrywide data on snakebite of the period 1940–1949 and came to the conclusion that West Bengal (an eastern zone province) had the highest snakebite mortality cases. As is shown in Gaitonde and Bhattacharya (1980), the annual incidence of severe envenomation in Maharashtra (western zone province of India) is about 70 people per 100,000 inhabitants and the annual mortality rate is nearly 2.4 per 10,000 people. Jena and Sarangi in 1993 showed that India witnesses more than 300,000 cases of snakebite and about 10,000 snakebite deaths every year. According to Chippaux (1998), the incidence of snakebite in India ranges from 66 to 163 victims per 10,000 people, out of which 1.4–68 deaths occur every year. However, most snakebite reports in India are hardly based on field survey, so they are not fully reliable (Hati et al., 1992).

Another epidemiological survey on snakebite suggested that around 50,000 snakebite mortalities occur in India (Mohapatra et al., 2011). Majority of snakebites (97%) take place in rural areas, more specifically near or in agricultural fields. Therefore, snakebite is considered as an occupational health hazard for the rural people of the country (Mohapatra et al., 2011). Men as compared to women are prone to snakebite and snake envenomation is most common during the monsoon months from June to September (Mohapatra et al., 2011). This may be due to the fact that during this season, snakes come out from their dens either in search of food or due to submerging of their holes with rainwater.

A very recent study has shown that 1.2 million snakebite deaths (average 58,000/ year) have occurred in India from 2000 to 2019 (Suraweera et al., 2020). People in the age group of 30–69 years account for nearly half of the snakebite deaths and 25% of the snakebite victims are children below 15 years of age. Table 1.1 shows age-wise distribution of snakebite data in India covering a period from 2000 to 2019.

The snakebite survey also suggested that the annual snakebite deaths are the highest in the states of Uttar Pradesh, Odisha, Andhra Pradesh, and Bihar (Mohapatra et al., 2011). In another hospital-based statistical analysis of snakebite, it has been reported that northern India witnessed the highest annual mortality rate due to snakebite (Patil, 2013). In the state of Maharashtra (western India), an estimation showed 10,000 annual venomous snakebites that resulted in 2000 deaths (Warrell, 2010). Further, several actual field-based or hospital-based studies have

Table 1.1 Estimated snakebite deaths in thousands by age and sex from 2000 to 2019 in India

Age range	Male (LL, UL)	Female (LL, UL)	Both (LL, UL)
0–14 years	149 (134, 154)	176 (160, 180)	325 (294, 334)
15–29 years	109 (102, 111)	88 (82, 89)	197 (184, 199)
30–69 years	290 (269, 303)	253 (232, 260)	543 (501, 564)
70 years or above	54 (45, 60)	48 (44, 50)	102 (89, 110)
All ages	602 (551, 626)	565 (518, 578)	1167 (1068, 1204)

LL Lower limit, *UL* Upper limit. These are lower and upper uncertainty bounds for estimates
Adopted from Suraweera et al. (2020)

suggested that the states of West Bengal, Tamil Nadu, Chhattisgarh, and Kerala also have the extraordinary incidence of snakebite cases (Mohapatra et al., 2011; Philip, 1986). Notwithstanding the fact that it is one of the mega biodiversity hot spots of the world containing a wide variety of venomous snakes, for example *N. kaouthia* and *B. fasciatus*, snakebite mortality and morbidity have hardly been reported from the north-eastern states of India, which warrants a detailed epidemiological study of snakebite in these regions. Furthermore, sea snake envenomation in the coastal areas of India is also unreported because this snake usually does not come in contact with humans (fishermen) or due to lack of awareness, such bites are not reported. A recent epidemiology of snakebite has shown the spatial distribution of snakebite mortality risk in India from 2004 to 2013 (Fig. 1.5). Snakebite death rates in different states of India are shown in Table 1.2.

The types of snake species prevalent in a particular geographical location of the country also influence the predominance of specific snakebite in that region. However, due to lack of a proper coordinated survey on snakebite, poor maintenance of hospital records, and dearth of awareness among people, it is very difficult to envisage the type(s) of snakebite in a particular region of the country. Only some discrete hospital-based surveys have tried to shed light into the type and frequency of snakebite in a particular locality. For example, an investigation on snakebite incidences (from 2012 to 2014) reported in a tertiary health care of Kerala has shown that majority of bitten species of snakes (44%) were unidentified, which is a grave concern for efficient management of snakebite (Chandrakumar et al., 2016). The hump-nosed viper accounts for 8.8% of total snakebites whereas only 1% of the patients are bitten by Indian cobra and Indian common krait. The Indian Russell's viper bite cases were 3.3% of the total snakebites during the study period. It has been reported that the frequency of snakebite in males (66.9%) surpassed that of female (34.1%) indicating the risk associated with outdoor occupational hazard (Chandrakumar et al., 2016). The report also shows that maximum incidence of snakebite takes place during afternoon (12–6 pm) and monsoon season (May–June). Another study over a period of 10 years (2000–2009), conducted in a single hospital in Kerala, shows that out of 1051 snakebite patients admitted to that hospital, 56% of victims showed hemotoxic symptoms typical of Viperidae bites, whereas 44% of patients displayed neurotoxic symptom due to elapid snakebite (Menon et al., 2016).

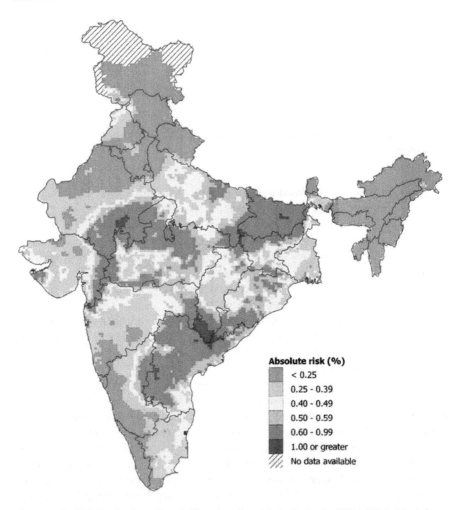

Fig. 1.5 Spatial distribution of snakebite mortality risk in India for 2004–2013 (cited from Suraweera et al., 2020)

A household survey among 28,494 people in the interior rural villages in Tamil Nadu state indicates that snakebite has a substantial and inconsistent influence on rural populations (Vaiyapuri et al., 2013). The study has shown that 3.9% of the people had been bitten by a snake and 20% of these had been bitten more than once in their lifetime and these envenomations result in 0.45% death. Males were more susceptible to bite as compared to females and the majority of bites occurred on the lower limbs. Russell's viper bite followed by cobra bite predominated in this region and frequencies of saw-scaled viper and krait bite were comparatively less. However, a major proportion of bitten species of snake could not be identified by patients or treating clinicians (Vaiyapuri et al., 2013). Annual snakebite frequency and rainfall statistics for Tamil Nadu state of India covering a period from 2001 to

Table 1.2 Data showing snakebite death rates in different states of India from 2001 to 2014

States of India	Study deaths in MDS	Annual average standardized death rate/100,000			Estimated deaths for 2001–2014 (000)
		2001–2004	2005–2009	2010–2014	
Higher burden states	1726	5.9	6.1	6.2	557.4
Andhra Pradesh	271	8.5	7.3	5.6	82.9
Bihar	321	5.6	7.6	8.9	101.9
Odisha	191	7.5	7.2	5.9	40.3
Madhya Pradesh	195	6.7	7.7	6.0	67.8
Uttar Pradesh	322	5.2	5.9	6.0	153.6
Rajasthan	192	4.9	6.7	5.0	52.1
Gujarat	176	4.1	4.8	5.1	38.8
Jharkhand	58	4.9	2.0	7.1	20.1
Lower burden states	1107	3.7	3.1	2.1	249.9
Chhattisgarh	42	6.0	6.5	2.5	16.8
Jammu and Kashmir	64	5.3	7.0	0.9	7.0
Tamil Nadu	176	6.1	3.4	3.0	42.1
Karnataka	137	5.6	3.3	2.9	33.0
Maharashtra	147	4.2	3.7	2.6	56.0
West Bengal	188	4.1	3.3	2.9	42.7
Punjab	67	2.9	3.1	4.0	14.5
Haryana	45	2.9	3.3	1.8	9.5
Assam	27	2.8	0.7	2.1	7.3
North-eastern states	37	2.3	0.9	0.7	2.4
Kerala	43	1.8	1.3	0.5	6.5
All other states	134	4.3	3.9	3.2	12.2
All India	2833	5.1	4.9	4.5	807.5

Adopted from Suraweera et al. (2020)

2010 showed that September–November months were the peak season of snakebite. Similarly, in another study on snakebite a tertiary care center (Sri Chamarajendra District Hospital) in southern India showed that out of the total 180 cases of snakebite, 108 patients were bitten by viper showing hemotoxic manifestations and 74 elapid-bite (cobra and krait) patients presented neuroparalytic symptoms (Halesha et al., 2013) suggesting that viper bite predominates in southern India region (Vaiyapuri et al., 2013; Halesha et al., 2013). However, these limited hospital-based epidemiological studies may not be sufficient to draw a firm conclusion on snakebite in this region.

In a hospital-based snakebite epidemiological survey, a total of 633 snakebite victims admitted to the Rural Community Centre and Punde Hospital in Mukhed taluka, Nanded district (Marathwada) of Maharashtra state, were analyzed retrospectively (Punde, 2005). It was estimated that 67.5% of the patients had been bitten by venomous snakes and 32.5% by nonvenomous snakes, and in between May and November maximum (67.5%) incidence of snakebite occurred. It was further reported that among the venomous snakebites, 64.2% of patients were bitten by saw-scaled viper, 16.6% by Indian cobra, 9.8% by Indian common krait, and 9.4% by Indian Russell's viper (Punde, 2005). Another study on snake envenoming conducted in rural Maharashtra showed that 30.2%, 26.3%, 22.5%, and 20.8% of patients were bitten by Indian saw-scaled viper, Indian common krait, Indian cobra, and Indian Russell's viper, respectively (Bawaskar et al., 2008). Snakebite envenoming in a tertiary care center in Maharashtra showed that Indian Russell's viper was the most common (9%) identified snake, followed by Indian common krait (5%); nonetheless, 20.3% of bites were by nonvenomous snakes (Padhiyar et al., 2018).

A study was undertaken in a tertiary care hospital in Central India (Chaudhari et al., 2014). It was observed that 63.4% of snakebites occurred during the rainy season (June–September); 75.8% and 23.8% of patients were bitten on lower and upper limbs, respectively; and mortality rate was 22.3%. Indian Russell's viper bite accounts for 28.5% of the total envenomation cases admitted to this tertiary care hospital, followed by Indian saw-scaled viper (11.5%), Indian common krait (8.1%), and Indian cobra bites (7.3%). Unidentified snakebites represented 44.6% of the total envenomations (Chaudhari et al., 2014). In Rajasthan, located in the northwestern part of the country, systemic envenoming due to *E. sochureki* is a severe problem; however, antidote against the venom of this species of snake is not included in commercial polyantivenom, which is a grave concern for the effective management of snakebite in this region (Kochar et al., 2007).

A study on the clinico-epidemiological profile and the treatment consequence of the snakebite cases from a secondary care center of Jharkhand, North India, shows that from January 2007 to December 2012, most of the snakebites occur during the monsoon season, July and August (Mitra et al., 2015). Indian krait bite (33.9%) followed by Indian Russell's viper bite (25.7%) was the most frequent; the incidence of Indian cobra bite was only 3.1% (Mitra et al., 2015). Nevertheless, majority of snakebites remained unidentified. The foot (48.3%) and hand (24.2%) were identified as the most common sites of bite (Mitra et al., 2015). Dandong et al. in 2018 also reported that deaths due to bite/sting of a venomous animal accounted for 10.7% in Bihar state of North India with an adjusted mortality rate of 6.2 per 100,000 population which was significantly higher in rural areas. Moreover, female mortality due to snakebite was found to be higher than in the males (Dandong et al., 2018). However, documentation on the type of bitten species of snake in Bihar is beyond the scope of this study and therefore not analyzed.

An epidemiological study on snakebite covering a period from 1 January 2013 through 31 December 2016, conducted in a tertiary health center (Ghatal Sub-divisional Hospital) of Paschim Medinipur district, West Bengal, has highlighted that out of 1160 snakebite cases, the mortality rate was only 1.6% (Manaa et al., 2019). About 82% of patients developed hemotoxic symptoms which is a typical feature of bites by Viperidae family of snakes. Because Indian Russell's viper predominates in this region, it may be well anticipated that this species of snake is responsible for major snakebites in this locale. About 11%, 2%, and 5% of the patients were bitten by *N. kaouthia, B. caeruleus*, and unidentified snakes, respectively (Manaa et al., 2019).

A hospital record-based epidemiological study of snakebites from 2008 to 2012 in Zonal Hospital, Solan, Himachal Pradesh, reported that 497 cases of snakebites were admitted during the study period (Gupt et al., 2015). Majority of snakebites (73%) were reported in the rainy season (July–September) and interestingly, the percent of females bitten by snakes (54%) was much more than males. A majority of patients showed hemotoxic symptom of snakebite and it was correlated with the prevalence of green pit vipers in this region, the venom of which is also known to be hemotoxic (Gupt et al., 2015). However, a study on the clinical profile of snake envenomation in a tertiary referral North Indian hospital from January 1997 to December 2001 showed that elapid bites were more than Viperidae bites with an overall mortality rate of 3.5% (Sharma et al., 2005). In another study conducted in a military hospital of North India, it was demonstrated that all of the envenomed patients displayed neurotoxic symptoms such as abdominal pain (91%), headache (86%), dysphagia (86%), ptosis (77%), diplopia (72%), blurring of vision (72%), dyspnea (67%), and vomiting (62%), all of which are typical clinicopathological manifestations of elapid bite (Singh et al., 2008). The median age of patients (all men) was 24 years and 81% of bites were in the lower third of leg and feet (Singh et al., 2008).

Very recently, Jamwal et al. (2019) conducted an epidemiological study to investigate the clinical profile and consequence of the neurotoxic envenomation in children in Jammu region and to identify the bitten species of snakes. The study has shown that both cobra and krait cause neurotoxic envenomation in children in Jammu region with krait bite accounting for 68% of total venomous bites and cobra bite accounting for 32% of the total cases (Jamwal et al., 2019). However, in 2004 it was reported that Indian saw-scaled viper envenomation is also a serious problem in the Jammu region of the Jammu and Kashmir state (Ali et al., 2004). Further, another Viperidae family of snake of medical significance, Levantine viper (*Macrovipera lebetina*), is also found in Jammu and Kashmir (Sharma et al., 2008). Envenomation by this species of snake shows necrosis and hemostatic manifestations in bite victims.

1.4   Key Issues Pertaining to Snakebite in India

In a nutshell, due to dearth of systematic protective measure at the national and public levels, India tops in the list of countries with maximum number of snakebite mortality and morbidity. Sadly, till date snakebite is not included under notifiable disease category in India. The snakebite epidemiological data from different regions of the country are fragmentary and mostly from hospital-based records. Several of the snakebite patients die on their way to the hospital; therefore, due to lack of systemic study it is very difficult to arrive at a concrete conclusion on the actual severity of envenomation in India and several of the other South and Southeast Asian countries where snakebite is a severe problem. However, the data have explicitly shown that the majority of snakebites occur in the rainy season, when the snakes come out from their dens in search of food and/or when the holes and pits are flooded with rainwater that results in an alarming enhancement of snake-human conflict. Therefore, people in general and health authorities in particular must be cautious during this period and there should be adequate facilities to treat snakebite patients including availability of antivenom in tertiary healthcare centers. Further, mostly males are the victims of snakebite compared to females because snakebite is an occupational health hazard and males are actively involved in field work or agricultural labor. Due to lack of door-to-door survey on snakebite and a well-coordinated herpetological survey, as well as nonexistence of a reliable snake envenomation detection kit, it is very difficult to say with certainty the distribution of particular species of venomous snakes across the country; nonetheless, the available data indicate that in southern India, Central India, western India, and eastern India bites by Viperidae family of snakes (Indian Russell's viper, Indian saw-scaled viper, and green pit viper in Himachal Pradesh) are more widespread whereas in northern India envenomation by the elapid family of snakes (Indian cobra and Indian common krait) is most common.

Notably, working-age-group persons (18–40 years) mostly from the rural areas are major victims of snakebite which is a foremost concern because many of them may be the only earning member of their family. This reflects the socioeconomic problem of the snakebite. Further, because the lower limbs are most vulnerable to snakebite, barefoot agricultural practice must be avoided and adopting proper precautions, for example wearing of gum boots and protective clothing during the monsoon season, can be of great help to prevent snakebite. The snakebite epidemiological reports also indicate that most of the bites take place in the evening. Hence, consequently possessing a flashlight particularly at dark may also save the situation and reduce the snakebite mortality. In the night, mosquito nets should be used more particularly when sleeping in floors. In addition to preventing snakebite this can also save from mosquito-borne diseases. In a nutshell, a combined effort from clinicians, toxinologists, antivenom manufacturing companies, and health authorities, along with awareness among mass, can certainly eradicate the snakebite problem.

References

Ali, G., Kak, M., Kumar, M., Bali, S. K., Tak, S. I., Hassan, G., & Wadhwa, M. B. (2004). Acute renal failure following *Echis carinatus* (saw-scaled viper) envenomation. *Indian Journal of Nephrology, 14*, 177–181.

Alirol, E., Sharma, S. K., Bawaskar, H. S., Kuch, U., & Chappuis, F. (2010). Snake bite in South Asia: A review. *PLoS Neglected Tropical Diseases, 4*(1), e603.

Bawaskar, H. S., Bawaskar, P. H., Punde, D. P., Inamdar, M. K., Dongare, R. B., & Bhoite, R. R. (2008). Profile of snakebite envenoming in rural Maharashtra, India. *Journal of the Association of Physicians of India, 56*, 88–95.

Chandrakumar, A., Suriyaprakash, T. N. K., LinuMohan, P., Thomas, L., & Vikas, P. V. (2016). Evaluation of demographic and clinical profile of snakebite casualties presented at a tertiary care hospital in Kerala. *Clinical Epidemiology and Global Health, 4*, 140–145.

Chaudhari, T. S., Patil, T. B., Paithankar, M. M., Gulhane, R. V., & Patil, M. B. (2014). Predictors of mortality in patients of poisonous snake bite: Experience from a tertiary care hospital in Central India. *International Journal of Critical Illness and Injury Science, 4*(2), 101–107.

Cheng, A. C., & Currie, B. J. (2004). Venomous snakebites worldwide with a focus on the Australia-Pacific region: Current management and controversies. *Journal of Intensive Care Medicine, 19*(5), 259–269.

Chippaux, J. P. (1998). Snakebites: Appraisal of the global situation. *Bull World Health Organ, 76*, 515–524.

Chippaux, J. P. (2011). Estimate of the burden of snakebites in sub-Saharan Africa: A meta-analytic approach. *Toxicon, 57*(4), 586–599.

Chippaux, J. P. (2012). Epidemiology of snakebites in Europe: A systematic review of the literature. *Toxicon, 59*, 86–99.

Chippaux, J. P., Bry, D., & Goyffon, M. (1995). Un type d'enquête sur les envenimations vipérines dans un département français: l'Yonne. *Bulletin de la Société herpétologique de France, 75/76*, 57–61.

Dandong, R., Kumar, G. A., Kharyal, A., George, S., Akbar, M., & Dandona, L. (2018). Mortality due to snakebite and other venomous animals in the Indian state of Bihar: Findings from a representative mortality study. *PLoS ONE, 13*(6), e0198900.

De Haro, L. (2012). Management of snakebite in France. *Toxcion, 60*, 712–718.

Fox, S., Rathuwithana, A. C., Kasturiratne, A., Lalloo, D. G., & de Silva, H. J. (2006). Underestimation of snakebite mortality by hospital statistics in the Monaragala District of Sri Lanka. *Transactions of the Royal Society of Tropical Medicine and Hygiene, 100*, 693–695.

Gaitonde, B. B., & Bhattacharya, S. (1980). An epidemiology survey of snakebite cases in India. *The Snake, 12*, 129–133.

Gold, B. S., Dart, R. C., & Barish, R. A. (2002). Bites of venomous snakes. *The New England Journal of Medicine, 347*(5), 347–356.

Gupt, A., Bhatnagar, T., & Murthy, B. N. (2015). Epidemiological profile and management of snakebite cases – A cross sectional study from Himachal Pradesh, India. *Clinical Epidemiology and Global Health, 3*, S96–S100.

Gutiérrez, J. M., Theakston, R. D. G., & Warrell, D. A. (2006). Confronting the neglected problem of snake bite envenoming: The need for a global partnership. *PLoS Medicine, 3*(6), e150.

Gutiérrez, J. M., Williams, D., Fan, H. W., & Warrell, D. A. (2010). Snakebite envenoming from a global perspective: Towards an integrated approach. *Toxicon, 56*, 1223–1235.

Habib, A. G., Kuznik, A., Hamza, M., Abdullahi, M. I., Chedi, B. A., Chippaux, J.-P., et al. (2015). Snakebite is underappreciated: Appraisal of burden from West Africa. *PLoS Neglected Tropical Diseases, 9*(9), e0004088.

Halesha, B. R., Harshavardhan, L., Lokesh, A. J., Channaveerappa, P. K., & Venkatesh, K. B. (2013). A study on the clinico-epidemiological profile and the outcome of snake bite victims in a tertiary care centre in Southern India. *Journal of Clinical and Diagnostic Research, 7*(1), 122–126.

Hati, A. K., Mandal, M., De, M. K., Mukherjee, H., & Hati, R. N. (1992). Epidemiology of snake bite in the district of Burdwan, West Bengal. *Journal of the Indian Medical Association, 90*, 145–147.

Hsiang, A. Y., Field, D. J., Webster, T. H., Behlke, A. D. B., Davis, M. B., Racicot, R. A., & Gauthier, J. A. (2015). The origin of snakes: revealing the ecology, behavior, and evolutionary history of early snakes using genomics, phenomics, and the fossil record. *BMC Evolutionary Biology, 15*, 87.

Hung, D.-Z. (2004). Taiwan's venomous snakebite: Epidemiological, evolution and geographic differences. *Royal Society of Tropical Medicine and Hygiene, 98*, 96–101.

Jamwal, A., Sharma, S. D., Saini, G., & Kour, T. (2019). Neurotoxic snakebite in Jammu Region: Is it cobra or krait. *International Journal of Scientific Study, 6*(11), 70–74.

Jena, I., & Sarangi, A. (1993). Snakebite. In *Snakes of medical importance and snake-bite management* (pp. 99–105). Ashish Publishing House.

Joseph, J. K., Simpson, I. D., Menon, N. C. S., Jose, M. P., Kulkarnia, K. J., Raghavendra, G. B., & Warrell, D. A. (2007). First authenticated cases of life-threatening envenoming by the hump-nosed pit viper (*Hypnale hypnale*) in India. *Transactions of the Royal Society of Tropical Medicine and Hygiene, 101*, 85–90.

Kalita, B., & Mukherjee, A. K. (2019). Recent advances in snake venom proteomics research in India: A new horizon to decipher the geographical variation in venom proteome composition and exploration of candidate drug prototypes. *Journal of Proteins and Proteomics, 10*, 149–164.

Kasturiratne, A., Wickremasinghe, A. R., de Silva, N., Gunawardena, N. K., Pathmeswaran, A., Premaratna, R., Savioli, L., Lalloo, D. G., & da Silva, H. J. (2008). The global burden of snakebite: A literature analysis and modeling based on regional estimates of envenoming and deaths. *PLoS Medicine, 5*, e218.

Kochar, D. K., Tanwar, P. D., Norris, R. L., Sabir, M., Nayak, K. C., Agrawal, T. D., Purohit, V. P., Kochar, A., & Simpson, I. D. (2007). Rediscovery of severe saw-scaled viper (*Echis sochureki*) envenoming in the Thar desert region of Rajasthan, India. *Wilderness & Environmental Medicine, 18*(2), 75–85.

Manaa, K., Ghosh, R., Gantaita, K., Sahaa, K., Paruaa, P., Chatterjee, U., & Sarkhel, S. (2019). Incidence and treatment of snakebites in West Bengal, India. *Toxicology Reports, 6*, 239–243.

Mathers, C. D., Ezzati, M., & Lopez, A. D. (2007). Measuring the burden of neglected tropical diseases: the global burden of disease framework. *PLoS Neglected Tropical Diseases, 1*(2), e114.

McNamee, D. (2001). Tackling venomous snake bites worldwide. *Lancet, 357*(9269), 1680.

Menon, J. C., Joseph, J. K., Jose, M. P., Dhananjaya, B. L., & Oommen, O. V. (2016). Clinical profile and laboratory parameters in 1051 victims of snakebite from a single centre in Kerala, South India. *Journal of the Association of Physicians of India, 64*(8), 22–29.

Mitra, S., Agarwal, A., Shubhankar, B. U., Masih, S., Krothapalli, V., Lee, B. M., Jeevan Kuruvilla, J., & Alex, R. (2015). Clinico-epidemiological profile of snake bites over 6-year period from a rural secondary care centre of Northern India: A Descriptive Study. *Toxicology International, 22* (1), 77–82.

Mohapatra, B., Warrell, D. A., Suraweera, W., Bhatia, P., Dhingra, N., Jotkar, R. M., Rodriguez, P. S., Mishra, K., Whitaker, R., & Jha, P. (2011). Million Death Study, C. Snakebite mortality in India: a nationally representative mortality survey. *PLoS Neglected Tropical Diseases, 5*(4), e1018.

Mukherjee, A. K., & Maity, C. R. (2002). Biochemical composition, lethality and pathophysiology of venom from two cobras--*Naja naja* and *N. kaouthia*. *Comparative Biochemistry and Physiology Part B, Biochemistry & Molecular Biology, 131*, 125–132.

Mukherjee, A. K., Kalita, B., Dutta, S., Patra, A., Maity, C. R., & Punde, D. (2021). Snake envenomation: Therapy and challenges in India. In S. P. Mackessy (Ed.), *Section V: Global approaches to envenomation and treatments, handbook of venoms and toxins of reptiles* (2nd ed.). CRC Press.

Padhiyar, R., Chavan, S., Dhampalwar, S., Trivedi, T., & Moulick, N. (2018). Snake bite envenomation in a tertiary care centre. *The Journal of the Association of Physicians of India, 66*(3), 55–59.

Parrish, H. M. (1966). Incidence of treated snakebites in the United States. *Public Health Reports, 81*(3), 269–276.

Patil, T. (2013). Snake bite envenomation: A neglected public health problem in India. *Medical Journal of Dr. D.Y. Patil University, 6*(2), 123–123.

Patra, A., Kalita, B., Chanda, A., & Mukherjee, A. K. (2017). Proteomics and antivenomics of *Echis carinatus carinatus* venom: Correlation with pharmacological properties and pathophysiology of envenomation. *Scientific Reports, 7*, 17119.

Philip, E. (1986). Snake-bites and scorpion sting. *Indian Pediatrics, 23*, 181–188.

Pillai, L. V., Ambike, D., Husainy, S., Khaire, A., Captain, A., & Kuch, U. (2012). Severe neurotoxic envenoming and cardiac complications after the bite of a 'Sind Krait' (*Bungarus cf. sindanus*) in Maharashtra, India. *Tropical Medicine and Health, 40*, 103–108.

Punde, D. P. (2005). Management of snake-bite in rural Maharashtra: a 10-year experience. *National Medical Journal of India, 18*(2), 71–75.

Rojas, G., Bogarin, G., & Gutierrez, J. M. (1997). Snakebite mortality in Costa Rica. *Toxicon, 35*, 1639–1643.

Russell, F. E. (1980). Snake venom poisoning in the United States. *Annual Review of Medicine, 31*, 247–259.

Sasa, M., & Vazquez, S. (2003). Snakebite envenomation in Costa Rica: A revision of incidence in the decade 1990–2000. *Toxicon, 41*, 19–22.

Senji Laxme, R. R., Khochare, S., de Souza, H. F., Ahuja, B., Suranse, V., Martin, G., Whitaker, R., & Sunagar, K. (2019). Beyond the 'big four': Venom profiling of the medically important yet neglected Indian snakes reveals disturbing antivenom deficiencies. *PLoS Neglected Tropical Diseases, 13*, e0007899.

Sharma, B. D. (1998). The venomous Indian snakes. In *Snakes in India: A Source Book* (pp. 115–124). Asiatic Publishing House.

Sharma, N., Chauhan, S., Faruqi, S., Bhat, P., & Varma, S. (2005). Snake envenomation in a north Indian hospital. *Emergency Medicine Journal, 22*(2), 118–120.

Sharma, S. K., Khanal, B., Pokhrel, P., Khan, A., & Koirala, A. (2003). Snakebite-reappraisal of the situation in Eastern Nepal. *Toxicon, 41*, 285–289.

Sharma, L. R., Lal, V., & Simpson, I. D. (2008). Snakes of medical significance in India: the first reported case of envenoming by the Levantine viper (*Macrovipera lebetina*). *Wilderness & Environmental Medicine, 19*(3), 195–198.

da Silva, C. J., Jorge, M. T., & Ribeiro, L. A. (2003). Epdemiology of snakebite in central region of Brazil. *Toxicon, 41*, 251–255.

Simpson, I. D., & Norris, R. L. (2007). Snakes of medical importance in India: Is the concept of the "Big 4" still relevant and useful? *Wilderness and Environmental Medicine, 18*, 2–9.

Singh, J., Bhoi, S., Gupta, V., & Goel, A. (2008). Clinical profile of venomous snake bites in north Indian military hospital. *Journal of Emergencies, Trauma, and Shock, 1*(2), 78–80.

Stahel, E., Wellauer, R., & Freyvogel, T. A. (1985). Vergiftungen durch einhei-mische Vipern (*Vipera berus* und *Vipera aspis*). Eine retrospektive Studie an 113 patienten. *Schweizerische Medizinische Wochenschrift, 115*, 890–896.

Steward, C. J. (2003). Snakebite in Australia: First aid and envenomation management. *Accident and Emergency Nursing, 11*, 106–111.

Suraweera, W., Warrell, D., Whitaker, R., Menon, G., Rodrigues, R., Fu, S. H., Begum, R., Sati, P., Piyasena, K., Bhatia, M., Brown, P., & Jha, P. (2020). Trends in snakebite deaths in India from 2000 to 2019 in a nationally representative mortality study. *eLife, 9*, e54076.

Swaroop, S., & Grab, B. (1954). Snakebite mortality in the world. *Bulletin of the World Health Organization, 10*, 35–76.

Vaiyapuri, S., Vaiyapuri, R., Ashokan, R., Ramasamy, K., Nattamaisundar, K., Jeyaraj, A., Chandran, V., Gajjeraman, P., Baksh, M. F., Gibbins, J. M., & Hutchinson, E. G. (2013). Snakebite and its socio-economic impact on the rural population of Tamil Nadu, India. *PLoS One, 8*(11), e80090.

Warrell, D. A. (1996). Clinical features of envenoming from snake bites. In C. Bon & M. Goyffon (Eds.), *Envenomings and their treatment, Lyon* (pp. 63–76). Foundation Marcel Merieux.

Warrell, D. A. (2010). Snake bite. *The Lancet, 375*(9708), 77–88.

Warrell, D. A., Gutierrez, J. M., Calvete, J. J., & Williams, D. (2013). New approaches & technologies of venomics to meet the challenge of human envenoming by snakebites in India. *The Indian Journal of Medical Research, 138*, 38–59.

Weinstein, S. A., Tamara, L. S., & Kardong, K. V. (2010). Reptile venom glands-form function and future. In S. P. Mackessy (Ed.), *Handbook of venoms and toxins of reptiles* (pp. 65–91). CRC Press.

Whitaker, R. (2006). *Common Indian snakes: A field guide.* Macmillan Indian Pvt. Ltd..

Whitaker, R., Captain, A., & Ahmed, F. (2004). *Snakes of India.* Draco Books.

World Health Organization. (1981). Progress in the characterization of venoms and standardization of antivenoms W.H.O. *WHO Offset Publication, 58*, 1–44.

World Health Organization. (2007). *Rabies and envenomings: a neglected public health issue: Report of a consultative meeting.* World Health Organization.

World Health Organization. (2011). Neglected tropical diseases: Snakebite, online publication. Retrieved from http://www.who.int/neglected_diseases/diseases/snakebites/en/201.

Evolution of Snakes and Systematics of the "Big Four" Venomous Snakes of India

2

Abstract

The evolution of snakes from an ancestral burrowing lizard remains disputed because the missing link between snakes and lizards is yet to be recognized. The available fossil record has led to the postulation that the ancestors of snakes made their first appearance during the Cretaceous Period. In recent decades, several studies have shown the mechanism of limb loss in snakes, though the concept is widely debated and the subject of speculation among evolutionary biologists. About 23–65 million years ago during two-thirds of the Tertiary Period, smaller python-like snakes were most prevalent on earth. Phylogenetic studies have indicated that the evolution of snakes involved a steady trend toward greater surface activity, enlargement of the body size, and an enlarged gape. The rear-fanged colubrid snakes evolved 35 to 55 million years ago; however, the origin of the Viperidae and Elapidae from the colubrid snakes is less clear. One group of researchers proposed a common single origin of the Viperidae and the Elapidae from colubrids while another group claimed that the elapids arose from the opisthoglyphous snakes and the viperids were derived from the proteroglyphous colubrids. In this chapter, different theories from evolutionary biologists are examined to explain the origin and evolution of snakes. The "Big Four" venomous snakes of India (*N. naja* and *B. caeruleus* of the Elapidae family, and *D. russelii* and *E. carinatus* of the Viperidae family) belong to the Infraorder Caenophidia and they all are members of the front-fanged advanced snakes. In addition, the occurrence of different species of the "Big Four" venomous snakes and their geographical distribution in the Indian subcontinent are discussed.

Keywords

Bungarus · Snake evolution · *Daboia* · *Echis* · Elapidae · Indian cobra · Indian krait · Indian Russell's viper · Indian saw-scaled viper · *Naja* · Origin of snakes · Systematics of the "Big Four" venomous snakes · Viperidae

2.1   Evolution of Snakes

The evolution of snakes is disputed and controversial. A large gap in our knowledge can be found with the initial evolution and phylogeny of snakes, which has consistently led to our state of uncertainty. The fossil record suggests that reptiles first appeared on earth some 300 million years ago during the Upper Carboniferous Period (Maris, 1997). Many reptiles including lizards are postulated to have emerged about 260–220 million years ago (Rieppel, 1988; Cundall & Irish, 2008). During the Cretaceous Period, the ancestors of snakes may have made their first appearance, and snake fossils of that time that were recovered from North Africa provide the evidence. Many scientists, using anatomical studies and phylogenetic analyses, have claimed that snakes evolved or developed from burrowing lizards (Rieppel, 1988; Cundall & Irish, 2008; Da Silva et al., 2018), though the "missing link" between snakes and lizards has not been clearly established (Maris, 1997). The squamate reptiles, the largest order of reptiles comprising lizards, snakes, and amphisbaenians (worm lizards that make up a group of legless squamates), represent one of the most specious clades (stemming from a common ancestor and all of its descendants both living and extinct) among terrestrial (living predominantly or entirely on land) vertebrate animals (Da Silva et al., 2018).

The central mystery is how, during evolution, did the snakes but not the lizards lose their limbs. Studies of the progression and development of the limbs and axial skeleton of snakes indicate that limb loss is correlated with body elongation (Head & Polly, 2015; Kvon et al., 2016). Nevertheless, evolutionary biologists have long speculated about the molecular basis for the loss of limbs in snakes and the topic continues to be unresolved (Zeller et al., 2009). Recently, Kvon et al. (2016) considered a series of recently sequenced snake genomes to analyze the molecular and functional evolution of a critical limb enhancer and to correlate its probable role in limb loss. Kvon et al. (2016) selected a limb-specific enhancer of the sonic hedgehog (Shh) gene, the zone of polarizing activity [ZPA] regulatory sequence (ZRS or MFCS1), which is considered as one of the best studied vertebrate enhancers in snakes (Lettice et al., 2014). The MFCS1 enhancer has been found to be activated in the posterior limb bud mesenchyme during the process of normal limb development in mouse (Sagai et al., 2005). A single-nucleotide mutation within the ZRS, which is highly conserved among vertebrates including fish but not in snakes, resulted in limb malformations. This suggested that it has a key role in limb development (Hill & Lettice, 2013). The ZRS enhancer gradually lost its function during the evolution of snakes and its loss in function has been shown to be associated with snake-specific nucleotide changes that may have contributed to the morphological disappearance of snake limbs through evolution (Kvon et al., 2016). The authors suggested that, in addition to ZRS enhancer changes in other sequences involving limb development (e.g., Hox genes that act upstream of Shh or other genes that might be critical for commencing limb development), something similar may have also happened in snakes (Kvon et al., 2016).

Many of the snakes that existed during the Upper Jurassic Period are said to be similar in size to the present-day snakes. In any case, the smaller, python-like snakes dominated the earth during two-thirds of the Tertiary Period, from about 65 to 23 million years ago (Maris, 1997). The scolecophidians, representing miniature, burrowing, wormlike blind snakes, were considered the most basal clade of living snakes, followed by other small, burrowing taxa such as the pipe snakes and shieldtail snakes that have limited gapes (Scanlon & Lee, 2011). A transitional form of partly surface-active sunbeam snakes, such as Xenopeltis and Loxocemus, gradually evolved while the typical, mostly surface-active and large-gaped snakes (represented by the pythons, boas, and colubrids) are believed to lead to a single, derived radiation ("core macrostomatans") (Scanlon & Lee, 2011). Phylogenetic studies indicate that the evolution of snakes involved a steady trend toward greater surface activity, enlargement of body size, and enlarged gape (Scanlon & Lee, 2011). The rear-fanged colubrid snakes likely evolved from 55 to 35 million years ago, and during the Miocene Period, the elapid snakes first made their appearance from 23 to 21 million years ago (Maris, 1997). The origins of the Viperidae and Elapidae from the colubrid snakes are much less certain. According to one group of researchers, they had a common single origin while other authors claim that the elapids arose from the opisthoglyphous and the viperids were derived from proteroglyphous colubrids (Kardong, 1982).

Mitochondrial ribosomal DNA sequence-based barcoding has been used to derive the phylogenetic relationships between a representative elapid (*Naja Naja*), colubrine (*Coluber constrictor*), and viperid (*Vipera ammodytes*). The results show a close evolutionary relationship between the elapids and the colubrines, in contrast to the relationship with viperid snakes (Knight & Mindell, 1994). Thus, the phylogeny indicates that front-fanged venom systems evolved independently in elapids and viperids and researchers have theorized that a loss of the front fangs may have occurred in colubrines. This idea refutes the hypothesis that front-fanged venom systems evolved from a common origin and it also disproves a possible "sister status" for viperids and elapids (Knight & Mindell, 1994). The origin of the viperids is purported to have taken place before the appearance of the elapids, and the colubrine radiation and their ancestors may have resembled the Homalopsidae, an ancient group of typically stout-bodied water snakes with well-developed rear fangs that were mildly venomous (Knight & Mindell, 1994).

During the process of evolution, the pit vipers acquired heat-sensitive pits on the front of their face to detect warm-blooded prey during the night. Amazingly, these pits are sufficiently sensitive to detect the smallest difference in temperature, to one-thousandth of a degree. Another group of pit vipers, the rattlesnakes found in North America, developed a unique warning apparatus at the end of their tail. All venomous organisms, such as snakes, are armed with a venom gland with its primary function to synthesize and store a myriad of toxins to serve an important function for the snake. A detailed description of the composition and evolution of snake venom is described in the next chapter. Several theories to explain the evolution of snakes are based on experimental evidence. They are described in the following sections.

2.2   Studies of the Genomics, Phenomics, and Fossil Record Show the Origin and Evolution of Snakes

The ecological and evolutionary origins of the Pan-Serpentes (snake total group) and Serpentes (crown snakes) have long been debated (Hsiang et al., 2015). Hsiang and colleagues in 2015 used a combined-data approach with up-to-date information from the vestige record of defunct crown snakes, recent data on the anatomy of stem snakes (*Najash rionegrina*, *Dinilysia patagonica*, and *Coniophis precedens*), and a profound understanding of the distribution of phenotypic apomorphies among the major clades of fossil and recent snakes to present the first comprehensive analytical reconstruction of the ancestor of crown snakes and the ancestor of the snake total group. Their study has provided insights into when, where, and how snakes originated, and a comprehensive view of the early evolution of snakes until present (Hsiang et al., 2015).

For ancestral state reconstruction, 18,320 base pairs (bp) from 21 nuclear loci, 766 phenotypic characters, a mitochondrial locus, and 11 novel characters were considered (Hsiang et al., 2015). By including the ancestral state reconstructions, the authors demonstrated that the predecessors of the crown snakes and the total group snakes were characterized by nocturnal (active at night) and wide scavenging, though they may not have subdued their prey by constriction. Likely, they mostly relied on stealth when hunting. They once preyed on soft-bodied vertebrates and boneless invertebrates of sizes that were no larger than their head. They were land dwelling, and they preferred to stay in warm, well-watered, and well-vegetated locations. The snake total group may have originated on land during the middle Early Cretaceous Period, about 128.5 million years ago, with the crown group arising about 20 million years later, during the Albian stage (Hsiang et al., 2015). The ancestral snakes unambiguously originated on land, rather than in water, and by inference, the predecessor of the crown snakes originated on the Mesozoic supercontinent of Gondwana. Likely, the ancestor of the total group snakes ascended on Laurasia (Hsiang et al., 2015). The divergence time tree for the evolution of snakes, inferred from constrained topology, is shown in Fig. 2.1.

2.3   Studies on the Genomic Regression of Claw Keratin, Taste Receptors, and Light-Associated Genes and the Evolutionary Origin of Snakes

The evolutionary origin and diversification of lineages have contributed significantly to important modern traits; however, some studies have unambiguously pointed to regressive evolution (defined as the loss of previously adaptive traits during evolution) that can also significantly influence the shaping of clades and their phenotypic diversity (Albalat & Cañestro, 2016; Emerling, 2017). Phenotypic characters are known to retrogress in a lineage, even though their genomes would likely preserve a corresponding signature of the retention of unitary pseudogenes and whole-gene obliterations (Albalat & Cañestro, 2016; Emerling, 2017). Thus, evolutionary

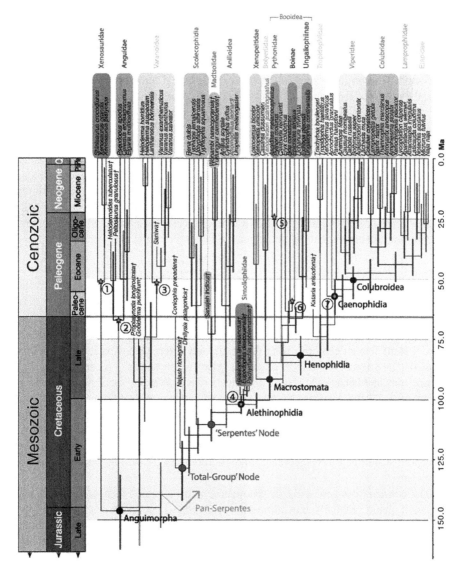

Fig. 2.1 Divergence time tree inferred from constrained topology (described by Hsiang et al., 2015). Divergence times were inferred from the constrained tree in BEAST. Major crown clades are named, with two extinct clades (Simoliophiidae and Madtsoiidae). The red line separating the Mesozoic and Cenozoic eras marks the Cretaceous-Paleogene (K-Pg) boundary at 66 Ma. The timescale is in millions of years (Ma). Circled numbers and green stars correspond to calibration dates outlined in the Additional file 12. Colored boxes indicate major clades. Fossil taxa are marked with a dagger (†). Grayed taxa names indicate extant species that are included based only on phenotypic data (the figure and its legend are adopted with permission from Fig. 8 of Hsiang et al., 2015)

biologists can test the different hypotheses of gene function to understand the evolutionary history by understanding the ancestries of phenotypes (Emerling, 2017). In 2017, Christopher Emerling from the University of California, Berkeley, USA, tested hypotheses of gene functionality, phenotype, and evolutionary antiquity to decipher the origin of snakes (Serpentes). He tested three hypotheses associated with the origin of snakes: (a) the claw-specific expression of keratins HA1 and HA2 genes; (b) the lack of taste buds in snakes; and (c) adaptation to dim-light conditions as a characteristic of the earliest snakes (Emerling, 2017).

Emerling (2017) found evidence in favor of hypotheses (a) and (b) but the retention of numerous taste receptors in some snakes led to his rejection of the hypothesis that snakes cannot taste their food or prey. The patterns of gene loss also inform the ecological origins of snakes, and some evidence also suggests that prior to limb loss, stem snakes were adapted to dim light and most of their loss of taste buds occurred in parallel within the crown lineages (Emerling, 2017). The temporal distribution of serpent and gekkotan gene losses is shown in Fig. 2.2.

2.4   Skull Evolution and the Ecological Origin of Snakes

Studies to understand the initial ecological and evolutionary origins of snakes were mainly concentrated on their distinct morphological differences, though the adaptive role of skull shape development in the origin and divergence of snakes was poorly studied (Da Silva et al., 2018). Different selective pressures (e.g., feeding performance, diet, and behavior) played a significant role in affecting the skull morphology (Fabre et al., 2016). Nevertheless, the relationship between development and natural selection in driving the morphological skull discrepancies during the lizard-to-snake transition is not well understood. To elucidate the mystery of the ecological and evolutionary origins of snakes, Da Silva et al. (2018) performed a series of analyses in a large-scale integrative geometric morphometric study of skull bones across the squamates.

The comparative geometric morphometric study revealed that during evolution the skull shape gradually transitioned from lizards to snakes, but not through the entire estimated morphospace (Da Silva et al., 2018). The study also suggested that the transition from lizards to snakes could have taken place through fossoriality, rather than by conversion that followed some other evolutionary path. The authors explicitly stated that the evolution of the snake skull is a sound example of the balance between temporal regulation of morphogenesis (heterochrony) and natural selection (ecology) (Da Silva et al., 2018).

2.5   Systematics of the "Big Four" Venomous Snakes of India

The association between venomous snake systematics and toxinology is another subject of debate. With recent advances in tools, techniques, and molecular methods for identifying taxonomies (e.g., nuclear DNA phylogeny), new evidence can help

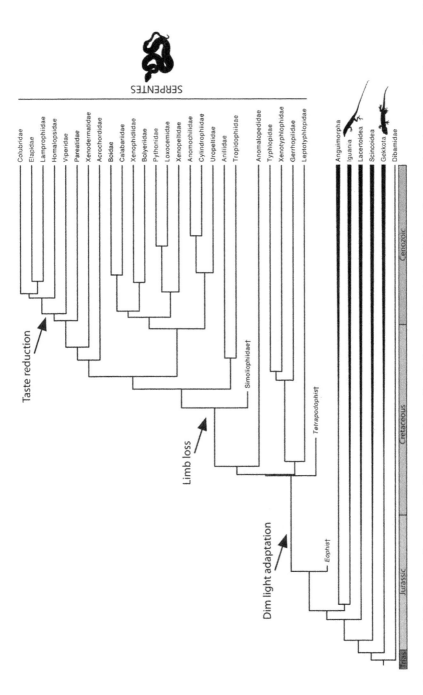

Fig. 2.2   Temporal distribution of serpent and gekkotan gene losses. Bars indicate gene inactivation; dating estimates are based on the means from four model assumptions (reprinted with permission from Fig. 1 of Emerling, 2017)

clarify the many hypotheses on snake taxonomy and systematics. Systematics has become one of the fastest growing and challenging subjects of the decade. During the twentieth century, the traditional classification of snakes was based on morphology, and at the beginning of the twenty-first century, snake systematics, guided by their morphological features, was also using molecular data (Burbrink & Crother, 2011). The relative and evolutionary analysis of molecular data (i.e., molecular systematics) offers a commanding statistical basis for hypothesis testing and assessing evolutionary progressions, which has gained a significant momentum in the modern systematics of venomous snakes. The diverse molecular dataset includes immunological distance (Cadle, 1988), allozyme electrophoresis (Dessauer et al., 1987), and DNA sequencing of portions of the mitochondrial small and large sub-unit ribosomal RNA genes (Knight & Mindell, 1994). The major benefit of studying molecular systematics is that, in addition to providing thousands to millions of characters, it also reveals the species tree relationships from the independently evolving gene estimates that are free from linkage or convergence (Burbrink & Crother, 2011). The modern method of venomous snake systematics has played a profound role in examining the relationships among the snakes at the species, genus, and family level. Nevertheless, traditional snake systematics continues to be significant and just a few molecular markers may contribute significantly to changing our understanding of snake phylogenomics and coalescent-based species tree estimations (Edwards, 2009; Burbrink & Crother, 2011).

Some of the latest discoveries have led to drastic changes in cataloging several groups of medically important venomous snakes that may change our concept of snake taxonomy and the origin of snakes (Wüster et al., 1997). Further, a significant decrease is expected to occur in the cost of next-generation DNA sequencing, which will add valuable data for modern snake systematics and prompt changes to our concepts of snake taxonomy and their evolutionary relationships. Because this chapter is dedicated to studies of the systematics of the "Big Four" venomous snakes of India, details about the classification of other snakes are beyond the scope of this discussion.

2.6   The "Big Four" Venomous Snakes of India Represent the Advanced Group of Snakes

The "Big Four" venomous snakes of India (*N. naja*, *B. caeruleus*, *D. russelii*, and *E. carinatus*) are members of the front-fang advanced snakes, belonging to Infraorder Caenophidia, which is comprised of approximately 80 snake species of the world (Wüster et al., 1997; Keogh, 1998; Burbrink & Crother, 2011). The Caenophidia includes four major groups: (a) the primarily nonvenomous colubrids; (b) the venomous atractaspids, and two independently evolved venomous groups; (c) the Viperidae (vipers, rattlesnakes, and their allies); and (d) the elapids (coral snakes, cobras, sea snakes, and their allies) (Wüster et al., 1997; Keogh, 1998; Burbrink & Crother, 2011). The Indian cobra and Indian common krait belong to the elapids, whereas Indian Russell's viper and Indian saw-scaled viper are included in the

Viperidae (McCarthy, 1985; Dessauer et al., 1987). The Elapidae and Viperidae are monophyletic; the fangs of the former group are "fixed" (more or less immovable), whereas the fangs of the viperids are anchored to a rotating maxilla that allows the bigger length fangs to keep folded in the mouth when not in use. They are erected at the time of venom injection to the snake's prey or victim, just after opening the mouth (Knight & Mindell, 1994). The following sections present a brief description of each family of the "Big Four" snakes.

2.6.1   Family Elapidae

The elapids are placed in the family Elapidae or families Elapidae and Hydrophiidae (Keogh, 1998). Elapids have the unique characteristic of possessing two enduringly upright canaliculated front fangs (termed the proteroglyphous condition) (McCarthy, 1985). Elapids are comprised of approximately 300 species and are spread across the tropical and subtropical countries including the Americas, Africa, Asia, Melanesia, Australia, and the Indian and Pacific oceans (Golay et al., 1993; Keogh, 1998). The snakes in the Elapidae family are composed of six genera: (a) *Austrelaps*, (b) *Bungarus*, (c) *Disteira*, (d) *Micrurus and Leptomicrurus*, (e) *Naja*, and (f) *Pseudonaja* (Wüster et al., 1997; Keogh, 1998). For the purpose of this study, however, the description is limited to the genera *Bungarus* and *Naja*, two members of the "Big Four" venomous snakes of India.

2.6.1.1 Genus Bungarus (Kraits)
In 1994, Slowinski studied the phylogeny of the genus *Bungarus*, with its 15 species, some of which are the most venomous snakes of the world (Whitaker & Captain, 2004). The Indian common krait (*Bungarus caeruleus*) belongs to the genus *Bungarus* and it is distributed across the Indian subcontinent including Sri Lanka, eastern Pakistan, and Southeast Asia (Quijada-Mascarenas & Wuster, 2010). The different species of *Bungarus* in the Indian subcontinent are described in Table 2.1.

2.6.1.2 Genus Naja (Cobras)
The genus *Naja* is comprised of approximately 38 species, but based on their morphological features and molecular systematics, the taxonomy of Asiatic species of *Naja* has been the subject of extensive revision (Wüster & Thorpe, 1989; Wüster et al., 1995, 1997, 2007). In the Indian subcontinent (India, Pakistan, Bangladesh, Sri Lanka, and Nepal) *N. naja* is more prevalent, while a species of the Indian monocled cobra *N. kaouthia* is also prevalent in eastern and north-eastern parts of India where it is responsible for a large number of snakebite deaths (Mukherjee & Maity, 2002). In Thailand, the most common cobra is the Indochinese spitting cobra *N. siamensis*, though the varied colorations of this species have been confused with *N. kaouthia* (Wüster et al., 1995, 1997). In addition, *N. oxiana* has been reported in northwest India (Santra et al., 2019). The king cobra, which can raise its hood, is distinct from the genus *Naja*; and therefore, this species, which was an earlier homotypic synonym of *Naja hannah*, is taxonomically identified as *Ophiophagus*

Table 2.1 *Bungarus* species and their geographical distribution in the Indian subcontinent (Sharma, 1998; Quijada-Mascarenas & Wuster, 2010)

Species	Common name	Geographical distribution
B. andamanensis	South Andaman krait	India (Andaman Island)
B. bungaroides	Northeastern hill krait	Myanmar, India (Assam, Cachar, Sikkim), Nepal, Vietnam
B. caeruleus	Common krait, Indian krait	Afghanistan, Pakistan, India (Bengal, Maharashtra, Karnataka), Sri Lanka, Bangladesh, Nepal
B. ceylonicus	Ceylon krait, Sri Lankan krait	Sri Lanka
B. fasciatus	Banded krait	Bangladesh, Northeast India, Bhutan, Nepal
B. lividus	Lesser black krait	India, Bangladesh, Nepal
B. niger	Black krait	India (Assam, Sikkim), Nepal, Bangladesh, Bhutan
B. sindanus	Sind krait	Southeast Pakistan, India
B. walli	None	India (Uttar Pradesh), Nepal, Bangladesh

Table 2.2 *Naja* species and their geographical distribution in the Indian subcontinent (Sharma, 1998; Whitaker & Captain, 2004; Santra et al., 2019)

Species	Common name	Geographical distribution
N. naja	Common cobra, Indian cobra, spectacled cobra	Throughout India (however its presence in Upper Assam and other parts of NE India is very less), Pakistan, Bangladesh, Bhutan, Nepal, Sri Lanka
N. kaouthia	Monocellate, monocled cobra, Bengal cobra	Bengal, Odisha, Assam, Andaman Islands
N. oxiana	Black cobra, acellate cobra, Oxus cobra	Northwest India, Northern Pakistan
N. sagittifera	Andaman cobra	Andaman Islands of India

hannah (Sharma, 1998). The king cobra is widely found in various parts of India, including Western Ghats, South of Goa, Andaman Islands, Himalayas, Bihar, Sundarbans region of West Bengal, and Odisha (Sharma, 1998; Whitaker & Captain, 2004). The variety of prevalent species (genus *Naja*) in India are shown in Table 2.2.

2.6.2 Family Viperidae

The snakes of the family Viperidae (vipers) are distributed in most parts of the world, except in Antarctica, Australia, Hawaii, Madagascar, and New Zealand. The Viperidae possess comparatively long solenoglyphous (hollow) fangs that allow deep penetration and injection of venom to its prey or victim (Knight & Mindell, 1994). The family Viperidae is divided into three subfamilies: (a) Azemiopinae, (b) Crotalinae, and (c) Viperinae (Knight & Mindell, 1994; Quijada-Mascarenas & Wuster, 2010). The subfamily Viperinae is comprised of the genera *Daboia*,

Macrovipera, and *Vipera*. Russell's viper and saw-scaled viper of the "Big Four" snakes belong to the subfamily Viperinae, which are also widespread in Europe, Asia, and Africa.

2.6.2.1 Genus Daboia

In 1983, Obst proposed that several hitherto known species of viper (*Vipera russelii*, *V. lebetina*, *V. mauritanica*, *V. palaestinae*, and *V. xanthina* and allied species) should be revalidated and merged with the genus *Daboia* (Obst, 1983). Almost 10 years later, Herrmann et al. (1992) used an albumin immunological method to re-examine the phylogeny of the genus *Vipera*. Their findings supported the revalidation of *Daboia* as a monotypic genus (including Russell's viper (RV), now *Daboia russelii*) (Lenk et al., 2001). Based on morphological variations (differences in color and markings) Wüster and Thorpe (1992) suggested that RV consists of two highly distinct populations: (a) the western part (India, Pakistan, Sri Lanka, and Bangladesh) and (b) the eastern part (China, Taiwan, Thailand, Burma, and Indonesia). RV was subsequently classified into five subspecies: *Daboia russelii russelii* (India, Pakistan, Nepal, and Bangladesh), *Daboia russelii pulchella* (Sri Lanka), *Daboia russelii siamensis* (Thailand, Myanmar, and China), *Daboia russelii formosensis* (Taiwan), and *Daboia russelii limitis* (Indonesia) (Wüster & Thorpe, 1992).

In 1996, however, Tsai et al. suggested the occurrence of two types of RV based on the presence of either asparagine (Asn, N) or serine (Ser, S) at the N-terminus of the phospholipase A_2 (PLA_2) isoenzymes of the venom (Tsai et al., 1996). Although taxonomic classifications are rarely made on the basis of a single enzyme in the venom, using morphological characteristics coupled with a mitochondrial DNA analysis, RV has been classified into two species: *D. russelii*, inhabiting the Indian subcontinent (but far less prevalent in Assam and other parts of Northeast India), and *D. siamensis*, which is prevalent in parts of Southeast Asia (other than the Indian subcontinent), southern China, and Taiwan (Thorpe et al., 2007). Despite the massive venom variation of RV in different geographical regions (Kalita & Mukherjee, 2019), the differences do not fully correspond to either subspecies recognized by Wüster and Thorpe (1992) or the conventional subspecies recognized previously (Wüster et al., 1997). Thus, the molecular taxonomy of RV across the country needs to be carefully examined.

2.6.2.2 Genus Echis

Echis spp. (common names: saw-scaled viper or carpet viper) are prevalent in the dry regions of Africa, the Middle East, and the Indian subcontinent. While this genus is comprised of approximately 12 full species, *E. carinatus* (also known as the saw-scaled viper) is widespread in southern India, western India, Sri Lanka, and Pakistan. Despite considerable progress being made in identifying these snakes, classification of the saw-scaled or carpet vipers of the genus *Echis* remains problematic (Wüster et al., 1997). Until the 1980s, only two species of *Echis* had been documented: (a) *Echis coloratus*, prevalent in Egypt, Israel, and the Arabian Peninsula, and (b) *Echis carinatus*, comprising the inhabitants from West Africa to India and Sri Lanka (Wüster et al., 1997). *E. carinatus* has since been divided into a number of subspecies, though many classifications have often been inconsistent (Wüster et al., 1997). A considerable debate exists on whether the taxa *sochureki* and *multisquamatus* would be considered as full species or subspecies of Asian *Echis carinatus* (Wüster et al., 1997). According to another classification scheme, *Echis carinatus* is divided into two subspecies: *E. c. sochureki* and *E. c. carinatus*, with the latter reported to be confined to peninsular India (Cherlin & Hughes, 1984).

In a nutshell, today, the analysis of *Echis* systematics is still inadequate and subject to debate. Nevertheless, appropriate coordination among clinicians, toxinologists, and herpetologists is necessary to acquire and disseminate accurate information about this genus for better hospital management of *Echis* envenomation.

References

Albalat, R., & Cañestro, C. (2016). Evolution by gene loss. *Nature Reviews. Genetics, 17*, 379–391.

Burbrink, F. T., & Crother, B. I. (2011). Evolution and taxonomy of snakes. In R. D. Aldridge & D. M. Sever (Eds.), *Reproductive biology and phylogeny of snakes* (pp. 19–53). CRC Press.

Cadle, J. E. (1988). *Phylogenetic relationships among advanced snakes. A molecular perspective* (Vol. 119, pp. 1–77). University of California Press.

Cherlin, V. A., & Hughes, B. (1984). New facts on the taxonomy of snakes of the genus *Echis*. *Smithsonian Herpetological Information Service, 61*, 1–7.

Cundall, D., & Irish, F. (2008). In C. Gans et al. (Eds.), *Biology of the Reptilia* (Vol. 20, pp. 349–692). New York Society for the Study of Amphibians and Reptiles.

Da Silva, F. O., Fabre, A.-C., Savriama, Y., Ollonen, J., Mahlow, K., Herrel, A., & Nicolas Di-Poï, J. M. (2018). The ecological origins of snakes as revealed by skull evolution. *Nature Communication, 9*, 376.

Dessauer, H. C. J., Cadle, E., & Lswson, R. (1987). Patterns of snake evolution suggested by their proteins. *Fieldiana: Zoology.*, new series, *34*, 1–34.

Edwards, S. V. (2009). Is a new and general theory of molecular systematics emerging? *Evolution, 63*, 1–19.

Emerling, C. A. (2017). Genomic regression of claw keratin, taste receptor and light-associated genes provides insights into biology and evolutionary origins of snakes. *Molecular Phylogenetics and Evolution, 115*, 40–49.

Fabre, A. C., Bickford, D., Segall, M., & Herrel, A. (2016). The impact of diet, habitat use, and behavior on head shape evolution in homalopsid snakes. *Biological Journal of the Linnean Society, 118*, 634–647.

Golay, P., Smith, H. M., Broadley, D. G., Dixon, J. R., McCarthy, C. J., Rage, J.-C., Schitti, B., & Toriba, M. (1993). *Endoglyphs and other major venomous snakes of the world: A checklist*. Azemiops S.A. Herpetological Data Center.

Herrmann, H.-W., Joger, U., & Nilson, G. (1992). Phylogeny and systematics of Viperinae snakes. III: Resurrection of the genus Macrovipera (Reuss, 1927) as suggested by biochemical evidence. *Amphibia-Reptilia, 13*(4), 375–392.

Hill, R. E., & Lettice, L. A. (2013). Alterations to the remote control of Shh gene expression cause congenital abnormalities. *Philosophical Transactions of the Royal Society of London. Series B, Biological Sciences, 368*, 20120357.

Hsiang, A. Y., Field, D. J., Webster, T. H., Behlke, A. D. B., Davis, M. B., Racicot, R. A., & Gauthier, J. A. (2015). The origin of snakes: revealing the ecology, behavior, and evolutionary history of early snakes using genomics, phenomics, and the fossil record. *BMC Evolutionary Biology, 15*, 87.

Head, J. J., & Polly, P. D. (2015). Evolution of the snake body form reveals homoplasy in amniote Hox gene function. *Nature, 520*, 86–89.

Kalita, B., & Mukherjee, A. K. (2019). Recent advances in snake venom proteomics research in India: a new horizon to decipher the geographical variation in venom proteome composition and exploration of candidate drug prototypes. *Journal of Proteins and Proteomics, 10*, 149–164.

Kardong, K. V. (1982). The evolution of the venom apparatus in snakes from colubrids to viperids and elapids. *Memórias do Instituto Butantan, 46*, 106–118.

Keogh, J. S. (1998). Molecular phylogeny of elapid snakes and a consideration of their biogeographic history. *Biological Journal of the Linnean Society, 63*, 177–203.

Knight, A., & Mindell, D. P. (1994). On the phylogenetic relationship of Colubrinae, Elapidae, and Viperidae and the evolution of front-fanged venom systems in snakes. *Copeia, 1*, 1–9.

Kvon, E. Z., Kamneva, O. K., Melo, U. S., Barozzi, I., Osterwalder, M., Mannion, B. J., Tissières, V., Pickle, C. S., Plajzer-Frick, I., Lee, E. A., Kato, M., Garvin, T. H., Akiyama, J. A., Afzal, V., Lopez-Rios, J., Rubin, E. M., Dickel, D. E., Pennacchio, L. A., & Visel, A. (2016). Progressive loss of function in a limb enhancer during snake evolution. *Cell, 167*, 633–642.

Lenk, P., Kalyabina, S., Wink, M., & Joger, U. (2001). Evolutionary relationships among the true vipers (Reptilia: Viperidae) inferred from mitochondrial DNA sequences. *Molecular Phylogenetics and Evolution, 19*(1), 94–104.

Lettice, L. A., Williamson, I., Devenney, P. S., Kilanowski, F., Dorin, J., & Hill, R. E. (2014). Development of five digits is controlled by a bipartite long-range cis-regulator. *Development, 141*, 1715–1725.

Maris, J. (1997). The origin of snakes. In *Snakes* (p. 10). Grang Books.

McCarthy, C. J. (1985). Monophyly of elapid snakes (Serpentes: Elapidae). An assessment of the evidence. *Zoological Journal of the Linnean Society, 83*, 79–93.

Mukherjee, A. K., & Maity, C. R. (2002). Biochemical composition, lethality and pathophysiology of venom from two cobras--*Naja naja* and *N. kaouthia*. *Comparative Biochemistry and Physiology Part B, Biochemistry & Molecular Biology, 131*, 125–132.

Obst, F. J. (1983). Zur Kenntnis des Schlangengattung *Vipera* (Reptilia, Serpentes, Viperidae). *Zoologische Abhandlungen, 38*, 229–235.

Quijada-Mascarenas, A., & Wuster, W. (2010). Recent advances in venomous snake systematics. In S. P. Mackessy (Ed.), *Handbook of venom and reptiles* (pp. 25–64). CRC Press.

Rieppel, O. A. (1988). Review of the origin of snakes. *Evolutionary Biology, 22*, 37–130.

Sagai, T., Hosoya, M., Mizushina, Y., Tamura, M., & Shiroishi, T. (2005). Elimination of a long-range cis-regulatory module causes complete loss of limb-specific Shh expression and truncation of the mouse limb. *Development, 132*, 797–803.

Santra, V., Owens, J. B., Graham, S., Wuster, W., Kuttalam, S., Bharti, U., Selvan, M., Mukherjee, M., & Malhotra, A. (2019). Confirmation of *Naja oxiana* in Himachal Pradesh, India. *Herpetological Bulletin, 150*, 26–28.

Scanlon, J. D., & Lee, M. S. Y. (2011). The major clades of living snakes: Morphological evolution, molecular phylogeny, and divergence dates. In R. D. Aldridge & D. M. Sever (Eds.), *Reproductive biology and phylogeny of snakes* (pp. 55–95). Science Publishers.

Sharma, B. D. (1998). Fauna of Indian snakes. In *Snakes in India: A source book* (pp. 87–108). Asiatic Publishing House.

Thorpe, R. S., Pook, C. E., & Malhotra, A. (2007). Phylogeography of the Russell's viper (*Daboia russelii*) complex in relation to variation in the colour pattern and symptoms of envenoming. *Herpetological Journal, 17*(4), 209–218.

Tsai, I. H., Lu, P. J., & Su, J. C. (1996). Two types of Russell's viper revealed by variation in phospholipases A_2 from venom of the subspecies. *Toxicon, 34*(1), 99–109.

Whitaker, R., & Captain, A. (2004). *Snakes of India: The field guide* (p. 495). Draco Books. ISBN 81-901873-0-9.

Wüster, W., & Thorpe, R. S. (1989). Population affinities of the Asiatic cobra (*Naja naja*) species complex in south-east Asia: Reliability and random resampling. *Biological Journal of the Linnean Society, 36*, 391–409.

Wüster, W., & Thorpe, R. S. (1992). Dentitional phenomena in cobras revisited: Spitting and fang structure in the Asiatic species of *Naja* (Serpentes: Elapidae). *Herpetologica, 48*(4), 424–434.

Wüster, W., Thorpe, R. S., Cox, M. J., Jintakune, P., & Nabhitabhata, J. (1995). Population systematics of the snake genus *Naja* (Reptilia: Serpentes: Elapidae) in Indochina: multivariate morphometrics and comparative mitochondrial DNA sequencing (cytochrome oxidase I). *Journal of Evolutionary Biology, 8*, 493–510.

Wüster, W., Crookes, S., Ineich, I., Mané, Y., Pook, C. E., Trape, J., & Broadley, D. G. (2007). The phylogeny of cobras inferred from mitochondrial DNA sequences: Evolution of venom spitting and the phylogeography of the African Spitting Cobras (Serpentes: Elapidae: *Naja nigricollis* Complex). *Molecular Phylogenetics and Evolution, 45*(2), 437–453.

Wüster, W., Philippe Golay, P., David, A., & Warrell, D. A. (1997). Synopsis of recent developments in venomous snake systematics. *Toxicon, 3*, 319–334.

Zeller, R., López-Rı'os, J., & Zuniga, A. (2009). Vertebrate limb bud development: Moving towards integrative analysis of organogenesis. *Nature Reviews Genetics, 10*, 845–858.

Snake Venom: Composition, Function, and Biomedical Applications

3

Abstract

The venom gland represents a modified salivary gland of venomous snakes for producing and storing venom toxins. Venom glands in the "Big Four" snakes, an advanced group of snakes, are situated in the temporal region behind the eye. The venom apparatus, which is considered as the evolutionary successor of Duvernoy's gland of colubrid snakes, is comprised of a highly specialized primary venom gland, a duct with an accessory gland, muscles for squeezing the venom, and fangs for delivering the toxic venom. The primary function of the venom is to immobilize the prey and also aid in the predigestion of prey. Evidence has also been presented for the involvement of specific components of venom in prey re-localization. The proteroglyph venom delivery system is found in the Indian cobra and Indian common krait, while the solenoglyph system occurs in Indian Russell's viper and Indian saw-scaled viper. This chapter presents a comprehensive review of the yield, lethal potency, and composition (enzymatic and nonenzymatic toxins) of the venom of the "Big Four" snakes. Further, the geographical and species-specific variation in snake venom composition and its impact on the clinical manifestation of snakebite in different regions of the country are discussed. The evolution of snake venom is highly debatable and several hypotheses, including the Toxicofera hypothesis and the independent origin hypothesis, have been proposed to explain the evolution of toxin genes in snake venom. These hypotheses, which include such ideas as the accelerated evolution of venom protein genes, selection pressure on the rapid adaptive evolution of venom proteins, and role of diet in snake venom evolution, have been proposed to explain the mechanism of evolution and diversification of snake venom toxins. The current scenario involving the development of drug prototypes from Indian snake venom toxins to treat various diseases is also presented to show the therapeutic potential of Indian snake venom proteins and polypeptides that may be explored to produce drugs to combat life-threatening diseases.

A. K. Mukherjee, *The 'Big Four' Snakes of India*,
https://doi.org/10.1007/978-981-16-2896-2_3

35

Keywords

Snake venom composition · Snake venom gland · Snake venom enzymes ·
Nonenzymatic toxins · Pathophysiology of snakebite · Antivenom treatment ·
Evolution of snake venom toxin genes · Functions of venom · Drug prototypes
from snake venom · Indian snakes

3.1   The Venom Gland and Venom Delivery Apparatus in the Viperidae and Elapidae Families of Snakes

The venom gland, a modified salivary gland of venomous snakes, produces and stores a mixture of toxins, known as snake venom. In advanced snakes (colubrids, viperids, elapids, and atractaspidids) the venom gland, which is situated in the temporal region behind the eye (along the upper jaw), evolved much later than in helodermatid lizards (Kardong, 1982; Kardong et al., 2009; Weinstein et al., 2009). Only a single major venom gland, a phylogenetic derivative of Duvernoy's gland, occurs in the Viperidae (Indian Russell's viper and Indian saw-scaled viper) and Elapids (Indian cobra and Indian common krait). The specialized venom apparatus is comprised of a highly specialized primary venom gland, a duct with an accessory gland, muscles for squeezing the venom, fangs for venom delivery, and toxic venom (Kochva, 1987; Jackson, 2003; Weinstein et al., 2009). During a strike by one of the "Big Four" venomous snakes, the striated muscles of the venom gland and the accessory gland are suddenly highly pressurized to deliver venom into the prey via a hollow fang (Weinstein et al., 2009).

Ancestors of the advanced snakes had long posterior maxillary teeth, but the teeth did not develop as fangs (Kardong, 1982). The fangs served as spikes that helped the snake grasp slippery, bulky, or tough prey during swallowing. As an aid to the rapid killing of agile prey, the fang and advanced venom gland system gradually evolved (Kardong, 1982). The morphological evolution of the venom gland, which resembled the evolution of fangs, commenced with the colubrid snakes (Kardong, 1982). Duvernoy's gland of the colubrid snakes is considered as the evolutionary predecessor of the venom gland of advanced snakes (Kochva, 1978, 1987). The series of morphological transformations of Duvernoy's gland to the venom gland of the Elapidae and Viperidae is shown in Fig. 3.1.

Venom glands are modified parotid salivary glands that contain three major cell types: (a) basal cells, (b) conical mitochondrion-rich cells, and (c) secretory cells (venom producing) (Minton, 1970). Despite having a similar basic design, the venom glands of the Indian cobra and the Indian common krait (Elapidae), and Indian Russell's viper and Indian saw-scaled viper (Viperidae), demonstrate some variability in their morphology and size. In the Indian Russell's viper and Indian saw-scaled viper, the venom is initially delivered from the main venom gland into the accessory gland by way of a primary duct. Then, from the accessory gland, the venom is passed into the base of the tubular fang via a secondary duct (Kardong et al., 2009) (Fig. 3.2). For the Indian cobra and Indian common krait, however, the

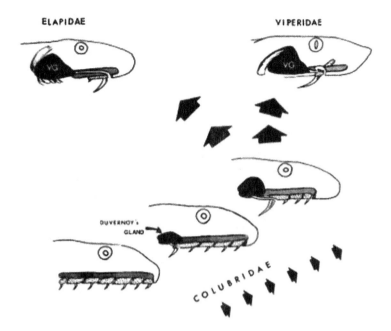

Fig. 3.1 Illustration of the independent transformation of Duvernoy's gland of Colubridae snakes, arising close to the posterior end of the supralabial gland (SLG), to become the advanced venom gland (VG) of the Viperidae and Elapidae (adopted from Kardong, 1982; sketch by Mr. Anandan Mukherjee)

accessory gland is located adjacent to the main venom gland and the venom is delivered to the accessory gland by way of the primary venom duct (Rosenberg, 1967; Weinstein et al., 2009; Kardong et al., 2009).

Two types of venom delivery (envenomation) systems are present in venomous snakes: (a) rear fanged and (b) front fanged (Vidal, 2002; Jackson, 2003; Weinstein et al., 2009). The former envenomation system (opisthoglyph) is characteristic of members of the Colubridae family (Mackessy & Saviola, 2016). The second (front-fanged) system is found in the Elapidae (Indian cobra and krait) and in the Viperidae (Indian Russell's viper and saw-scaled viper) family of snakes, which is further categorized into two types: (a) the proteroglyph system found in the Indian cobra and Indian common krait and (b) the solenoglyph system, which occurs in the Indian Russell's viper and Indian saw-scaled viper (Fig. 3.3). In the latter system, the long tubular fangs are placed on a small and exceedingly movable maxilla. The compressor muscles support the movements of the palatomaxillary arches during a strike by the Indian Russell's viper and as a result a large quantity of venom can be efficiently administered to the prey or victim (Deufela & Cundall, 2006).

The specific function of the accessory gland is debatable though it is believed to play a role in preventing unnecessary loss of venom toxins (Minton, 1970). Depending on the prey size, the accessory gland, which is under voluntary control, helps the snake to release an adequate amount of venom during envenomation

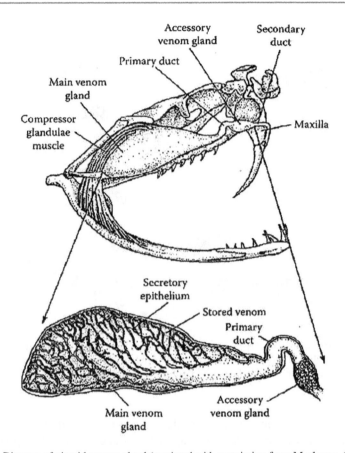

Fig. 3.2 Diagram of viperid venom gland (reprinted with permission from Mackessy, 1991)

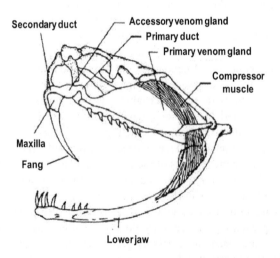

Fig. 3.3 Position of the venom gland with the accessory gland and their primary and secondary ducts, and fang in a viperid snake (reprinted with permission from Deufela & Cundall, 2006)

(Minton, 1970). Nevertheless, a proteomic analysis of the accessory venom gland secretion of *Bothrops jararaca* showed the occurrence of an ancillary source of toxins and enzymes (Valente et al., 2018).

During hibernation or long periods of fasting, venom is stored in the venom gland and no reports suggest that the stored venom toxins undergo a turnover (Mackessy & Baxter, 2006; Kardong et al., 2009). After extracting or milking venom, the secretory epithelium of the main venom gland immediately starts to synthesize venom proteins (toxins) (Mackessy, 1991; Kardong et al., 2009), with the complete synthesis of venom taking approximately 16 days (Kochva, 1987). After a natural strike by the snake, however, the venom is replenished more rapidly depending on the volume of the venom that was delivered to the prey or victim (Kochva, 1987; Weinstein et al., 2009).

3.2 Comprehensive Review of the Venom Composition of the "Big Four" Venomous Snakes of India

Snake venom is a modified saliva that developed during the evolution of snakes. Therefore, venom is an evolutionary adaptation to paralyze or immobilize swift prey and it is also used by snakes for defense (Weinstein et al., 2009). Venom is an extremely deadly secretion that is synthesized and subsequently stored in the specialized salivary glands of snakes, known as the venom gland. Venom is a unique mixture in terms of its biochemical and pharmacological properties, and it mostly consists of proteins and polypeptides (Kochva, 1987; Meier & Stocker, 1995). Most snake venoms are colorless, but some, such as the Indian Russell's viper and Indian saw-scaled viper venom, are yellowish as they contain L-amino acid oxidase (Mukherjee et al., 2015; Patra et al., 2017). The solid material in the venoms of the Elapidae (Indian cobra and Indian common krait) and Viperidae (Indian Russell's viper and Indian saw-scaled viper) ranges between 18 and 52%, and 28 and 31%, respectively (Elliott, 1978). Milked venom also contains some debris (Mukherjee, 1998; Mukherjee & Maity, 1998).

Snake venom is slightly acidic in nature and its specific gravity ranges from 1.03 to 1.07. The relative viscosity of snake venom varies from 1.5 to 2.5. The solubility of Elapidae venom in H_2O is much higher than the solubility of Viperidae venom, and the solubility of all venoms increases in physiological saline (Mukherjee, 1998). More than 90% of the snake venom components are mixtures of proteins and polypeptides, which accounts for their pharmacological and toxic effects in victims and animals (Stocker, 1990a, 1990b; Mukherjee, 1998). The proteins and polypeptides of snake venom can be further divided into enzymes and nonenzymes. Besides proteins, some nonprotein components like carbohydrates and metal ions are also present in snake venom, though they often occur in trivial amounts (Sect. 3.2.3).

The venom yield of the "Big Four" snakes varies depending on the species of snake. The specific venom yields (Murthy, 1998) and lethal doses of venoms (LD_{50} values) in experimental animals are shown in Table 3.1.

Table 3.1 Yield of venom and lethal potency of the "Big Four" venomous snakes of India

Snake species	Yield of venom from adult snake (mg)	Lethality (LD$_{50}$) in mice (mg/kg)	References
Indian cobra (*N. naja*)	150–275	0.4–1.2 0.2828 (i.v.)	Mukherjee and Maity (1998), Shashidharamurthy et al. (2010), Parveen et al. (2017)
Indian common krait (*B. caeruleus*)	8–12	0.325 (s.c.) 0.169 (i.v.) and 0.089 (i.p.) 0.2828 (i.v.)	Parveen et al. (2017), Wikipedia: Common Krait (https://en.wikipedia.org/wiki/Common_krait)
Indian Russell's viper (*D. russelii*)	130–250	2 to 5.4 (i.p.) 0.7 (i.v.) 0.3321 (i.v.)	Prasad et al. (1999), Mukherjee et al. (2000); Parveen et al. (2017)
Indian saw-scaled viper (*E. carinatus*)	8–12	0.5655 (i.v.)	Parveen et al. (2017), Wikipedia: *Echis carinatus* (https://en.wikipedia.org/wiki/Echis_carinatus)

The lethality (LD$_{50}$) values are determined by injecting the venom obtained from different geographical locations and using different routes of injection
sc subcutaneous, *iv* intravenous, *ip* intraperitoneal

3.2.1 Enzymatic Toxins of the "Big Four" Snake Venoms

The number of enzymes in snake venom can vary from venom to venom within the same species from different locales and for different species. Several of these enzymes have been identified by biochemical and proteomic analyses of the "Big Four" venomous snakes of India. Some have been purified to homogeneity and their biochemical and pharmacological properties have been characterized (reviewed by Kalita & Mukherjee, 2019). No single venom contains all of these enzymes, but some of the enzymes, like PLA$_2$, LAAO, proteases, and ATPase, are present in venom and the remaining enzymes are found to some extent in different species, including the "Big Four" venomous snakes of India (Stocker, 1990a, 1990b; Kang et al., 2011; Kalita & Mukherjee, 2019). Proteomics, transcriptomics, and a comparative study of enzyme activities have been used to biochemically analyze the venoms of 42 species belonging to the Colubridae, Elapidae, Viperidae, and Crotalidae snake families. The findings have led to the conclusion that the Elapidae (Indian cobra and Indian common krait) venoms are rich in phospholipases, phosphodiesterase, nucleotidase, ATPase, and cholinesterase, while the Indian Russell's viper and Indian saw-scaled viper venoms contain procoagulant and anticoagulant proteases, and kinin-releasing and arginine-ester-hydrolyzing enzymes (Stocker, 1990a, 1990b; Risch et al., 2009; Mukherjee, 1998, 2014a, 2014b; Kang et al., 2011; Mukherjee & Mackessy, 2013). Enzymes such as protease, acetylcholinesterase, ATPase, AMPase, L-amino oxidase, phospholipase A$_2$, and hyaluronidase in snake venom play an important role in inducing toxicity following snakebite (Table 3.1). Snake venom metalloproteinases (SVMPs), an abundant enzymatic protein in the Viperidae family of snakes, perform crucial roles in the adaptation of snakes to different ecological niches (Gutiérrez et al., 2009). Acetylcholinesterase, which is one of the toxic enzymes in cobra venom, acts on acetylthiocholine to liberate the choline and acetate, which may be responsible for the toxicity caused by Indian cobra bite (Guieu et al., 1994; Mukherjee & Maity, 2002). The $5'$-nucleotidase enzyme is responsible for hydrolysis of the terminal phosphate from adenylic acid (AMP). This is a Zn^{2+}- and EDTA-sensitive enzyme (Sales & Santoro, 2008). Venom ATPase, when injected, is known to "shock" victims due to the sudden hydrolysis of ATP (Kini & Gowda, 1982).

L-amino acid oxidase (LAAO) generates H_2O_2 to induce platelet aggregation and consequent thromboxane A_2 synthesis in the presence of Ca^{2+} but without ADP release (Li et al., 1994). Both procoagulant and anticoagulant enzymes, affecting the different stages of the blood coagulation cascade, have been purified and characterized from venoms of various snakes and from the "Big Four" snake venoms (Teng et al., 1984; Doley & Mukherjee, 2003; Sundell et al., 2003; Saikia et al., 2011, 2013; Mukherjee & Mackessy, 2013; Mukherjee, 2014a, 2014b).

A brief account of some of the major enzymes present in snake venom, especially the "Big Four" snakes, their mechanism of catalysis, and biological/pharmacological effects, is presented in Table 3.2.

3.2.2 Nonenzymatic Toxins in the "Big Four" Snake Venoms

Besides enzymes, snake venoms also contain numerous nonenzymatic proteins that play an important role in the toxicity of the venom, especially in immobilizing the prey before it is swallowed. In addition, structure-function characterizations of many of the nonenzymatic venom proteins (toxins) have led to the development of powerful research tools, diagnostic reagents, and invention of drugs and peptide-based therapeutic agents to combat various deadly diseases (see review by Mukherjee et al., 2011; McCleary & Kini, 2013). Various nonenzymatic proteins from snake venom, including the Indian "Big Four" snakes, have been isolated, purified, and characterized. Because their pharmacological activities have been difficult to neutralize by commercial antivenom, the hospital management of snake-bite patients has been complicated (Mukherjee & Maity, 2002; Kumar & Gowda, 2006; Mukherjee et al., 2014b). Neurotoxins isolated from snake venom have been found to impair nerve function, mainly by acting on neuromuscular transmission (Table 3.2). The neurotoxins can be further classified as (a) postsynaptically active neurotoxins, which block neuromuscular transmission by binding specifically to the AchE receptor, and (b) presynaptically acting neurotoxins that either inhibit transmitter release from nerve terminals or enhance the release of neurotransmitter (dendrotoxin) (Grant & Chiappinelli, 1985; Changeux, 1990; Kukhtina et al., 2000; Chung et al., 2002; Kini & Doley, 2010). The weak neurotoxin-like peptides (kaouthiotoxins) from *Naja kaouthia* venom interact with the phospholipase A_2 enzyme isolated from the same venom to synergistically enhance their cytotoxicity, as an example of nonenzymatic-enzymatic toxins that interact to enhance their

Table 3.2 List of enzymes identified in the Elapidae and Viperidae families of snakes according to their biochemical and/or proteomic analyses

Enzyme protein family	Molecular mass range (kDa)	Mechanism of action	Pharmacological effects	References
Phospholipase A_2 (PLA$_2$; EC 3.1.14)	13–18	Hydrolysis of sn-2 ester bond of phospholipids to release free fatty acids (FFA) and lysophospholipids, inhibition of thrombin and/or FXa of blood coagulation factors	Anticoagulation, myotoxicity, cytotoxicity, biomembrane damage, edema induction, hemolysis, neurotoxicity	Choumet et al. (1993), Doley and Mukherjee (2003), Doley et al. (2004), Mukherjee (2007), Mukherjee (2014a), Saikia et al. (2011, 2012, 2013), Mukherjee et al. (2014a), Dutta et al. (2015, 2019), Saikia and Mukherjee (2017)
Snake venom metalloprotease (SVMP; EC:3.4.24.42)	15–100	Activation of coagulation factor X and factor II, degradation of aα and/or bβ chains of fibrinogen and/or fibrin, binding and damage of capillary vessels, degradation of the subendothelial matrix proteins (fibronectin, laminin, type IV collagen, nidogen, and gelatins)	Myonecrosis, hemorrhage, blistering, necrosis, coagulopathy, inhibition of platelet aggregation, thrombocytopenia	Bjarnason and Fox (1994), Mukherjee (2008b), Chen et al. (2008), Thakur et al. (2015), Thakur and Mukherjee (2017a)
Snake venom serine protease (SVSP; EC: 3.4.21.-)	26–48	Presence of a typical and highly reactive "catalytic triad" in their active site, which comprises serine[195], histidine[57], and aspartic acid[102]; degradation of aα- and/or bβ-chains of fibrinogen and/or fibrin; and thrombin-like mechanism	Coagulopathy, fibrin(ogen)olytic (defibrinogenation of plasma), modulation of platelet aggregation	Mukherjee (2013, 2014b), Mukherjee and Mackessy (2013), Thakur and Mukherjee (2017a)

Enzyme	MW	Action	Effect	References
L-amino acid oxidase (LAAO; EC1.4.3.2)	50–70	v-LAAOs are flavoenzymes that catalyze oxidative deamination of L-amino acids to produce keto acids, ammonia, and hydrogen peroxide	Inhibition of platelet aggregation, coagulopathy, cytotoxicity (apoptosis induction)	Li et al. (1994), Du and Clemetson (2002), Ande et al. (2006), Fox (2013), Mukherjee et al. (2015), Ribeiro et al. (2016), Tan et al. (2018)
Trypsin-like esterase (TAME- or BTEE-esterase; EC 3.4.21.4)	26–48	Hydrolysis of aα and/or bβ chains of fibrinogen	Coagulopathy	Shimokawa and Takahashi (1995), Mukherjee (2014b), Kalita et al. (2017), Kalita, Patra, Das, and Mukherjee (2018a)
Exonuclease/ phosphodiesterase (PDE; EC3.1.4.1)	98–140	Phosphodiesterase enzyme breaks phosphodiester bonds in the 3′- to 5′-direction to yield nucleoside 5′-phosphates (exonuclease type a)	Inhibition of platelet aggregation, generation of purine nucleosides which play an important role in the overall pathophysiology of envenoming	Fox (2013), Kalita et al. (2017), Kalita, Patra, Das, and Mukherjee (2018a)
Endonuclease (i) DNase (EC 3.1.21.1)	~15	Cleavage of DNA	Generation of purine nucleosides which play an important role in the overall pathophysiology of envenoming	Sales and Santoro (2008), Fox (2013)
(ii) RNase (EC 3.1.21.-)		Cleavage of RNA		
Acetylcholinesterase (found in the Elapid family of snakes) (AchE; EC 3.1.1.7)	~65	Hydrolysis of acetyl choline (ach) to choline and an acetate group resulting in termination of the chemical impulse. The catalytic triad consists of Ser^{200}, Glu^{327}, and His^{440} residues	May be responsible for showing neurotoxicity or toxicity of cobra envenomation	Raba et al. (1979), Guieu et al. (1994), Frobert et al. (1997), Cousin et al. (1996), Ahmed et al. (2009)
Hyaluronidase (Hyal; EC 3.21.35)	33–110	Hydrolyzes a dermal barrier, the long-chain glycosaminoglycan hyaluronic acid	Spreading factor that helps invasion of venom by dissolution of extracellular matrix and connective tissues around the	Pukrittayakamee et al. (1988), Girish, Mohanakumari, et al. (2004a), Girish, Shashidharamurthy, et al.

(continued)

Table 3.2 (continued)

Enzyme protein family	Molecular mass range (kDa)	Mechanism of action	Pharmacological effects	References
			blood vessels, potentiation of local hemorrhage, and systemic envenomation	(2004b), Fox (2013), Vivas-Ruiz et al. (2019)
Nucleotidase: (i) ATPase (EC 3.6.1.3)	74–94	Hydrolysis of ATP to produce ADP/AMP and release inorganic phosphate	ATPase produces shock symptoms by the depletion of ATP and adds to prey immobilization	Zeller (1948), Zeller, 1950, Kini and Gowda (1982), Sales and Santoro (2008), Dhananjaya and D'Souza (2010), Kalita et al. (2017), Kalita, Patra, Das, and Mukherjee (2018a), Kalita, Patra, and Mukherjee (2018b), Patra et al. (2017)
(ii) ADPase (3.6.1.5)	>50	Hydrolysis of ADP to produce AMP and inorganic phosphate	Not known	
(iii) AMPase or 5'-nucleotidase (EC 3.1.3.5)	>50	Hydrolysis of phosphoester bond present in AMP and release of adenosine	Inhibition of ADP/collagen-induced platelet aggregation, generation of purine nucleosides which play an important role in the overall pathophysiology of envenoming. Released adenosine shows diverse pharmacological activities and could be involved in smooth muscle relaxation and vasodilation	
(iv) Apyrase (EC 3.6.1.5)	~80	Hydrolysis of phospho-anhydride bond in ATP and ADP but does not catalyze the cleavage of phosphoester bond present in AMP	Marginal anticoagulant, antiplatelet activity, contribution to pathophysiology of envenoming	Kalita, Patra, and Mukherjee (2018b)

Mechanisms of action and the biological activity of these enzymes are also shown overall toxicity (Mukherjee, 2008a). Cardiotoxins are responsible for cardiac arrest, muscle contracture, membrane depolarization, cytolysis, myonecrosis, and hemolysis; they affect platelets and also show bactericidal activity (Chen et al., 2007, 2011; Debnath et al., 2010).

Snake venom cytotoxins are low-molecular-weight toxic polypeptides responsible for inducing various pharmacological effects like hemolysis, cytolysis, depolarization of muscle membranes, and specific cardiotoxicity in prey or victims (Chen et al., 2007; Debnath et al., 2010; Méndez et al., 2011). Myotoxins contribute to the digestion of muscle cells and can cause significant skeletal muscle necrosis (Ownby, 1990). Protease inhibitors are low-molecular-mass Kunitz-type basic polypeptides of Elapidae and Viperidae venoms consisting of 52–65 amino acids and cross-linked by 2 or 3 disulfide bridges. They either act as proteinase inhibitors or represent structural analogs of proteinase inhibitors (Table 3.2). A tremendous scope exists for therapeutic applications of the Kunitz-type protease inhibitors isolated from snake venom.

Snake venom has also been reported to contain nerve growth factor (NGF) (Trummal et al., 2011), with its activity being identified in six Viperidae, nine Crotalidae, and five Elapidae species. It induces plasma extravasation and histamine release from whole-blood cells (Stocker, 1990a, 1990b; Trummal et al., 2011), with a mechanism that seems to differ from the mechanism of action of most commonly identified enzymes and toxins present in snake venom (Trummal et al., 2011). Although the exact mechanism of action of snake venom NGF is unknown, its cytotoxicity appears to be induced by an acidic phospholipase A_2 enzyme. The cognate complex from Indian cobra *N. naja* venom is significantly enhanced in the presence of trace amounts of NGF from the same venom (Dutta et al., 2019). A recent study has shown that NGF from Russell's viper venom can bind to the TrkA receptor expressed in breast cancer cells as well as in rat pheochromocytoma cell line (PC-12) derived from adrenal gland of *Rattus norvegicus* (Islam et al., 2020).

A summary of some major nonenzyme toxins present in snake venom, their mechanisms of action, and biological activity are shown in Table 3.3.

3.2.3 Nonprotein Components of Snake Venom

The nonprotein components of snake venom can be divided into two categories: (a) organic constituents and (b) inorganic constituents (Stocker, 1990a, 1990b). The organic constituents are carbohydrates (glycoproteins), lipids (phospholipids primarily), nucleosides and nucleotides, amino acids, biogenic amines (abundant in Viperidae and Crotalidae venoms) including histamine, serotonin, bufotenine, and N-methyl tryptophan. The inorganic constituents of snake venoms include Ca^{2+}, Fe^{2+}, Mn^{2+}, Na^+, Li^+, K^+, Co^{2+}, and Zn^{2+} and anions like phosphate, sulfate, and chloride. All of these substances are not found in every type of venom and the amount of each varies from species to species (Stocker, 1990a, 1990b).

Table 3.3 List of nonenzyme toxins identified in Elapidae and Viperidae families of snakes based on biochemical and/or proteomic analyses

Nonenzymatic toxin family	Molecular mass range (kDa)	Mechanism of action	Pharmacological activity	References
Three-finger toxins (elapid venom):	6–9			
(i) Cardiotoxins		Open the Ca^{2+}-release channel (ryanodine receptor located in the sarcoplasmic/endoplasmic reticulum membrane) and modify the activity of the Ca^{2+}-Mg^{2+}-ATPase in isolated sarcoplasmic reticulum preparations from cardiac or skeletal muscle β-Cardiotoxins are antagonists for the beta-1 and beta-2 adrenergic receptors	They show concentration-dependent pharmacological activity. Heart rate is increased at lower concentration of toxin. However, at higher concentrations they cause cardiac arrest of prey. Other pharmacological activities are hemolysis, cytolysis, contractures of muscle, membrane depolarization, and circulatory and respiratory failure. The β-cardiotoxins function as beta-blockers and decrease heart rate	Fletcher and Jiang (1993), Bilwes et al. (1994), Rajagopalan et al. (2007), Ponnappa et al. (2008), Debnath et al. (2010), Dutta et al. (2017)
(ii) Cytotoxin (CTx)		The basic nature of cytotoxins enables them to interact with anionic phospholipids that kill cells by non-selectively disrupting cell membranes	Hemolysis, cytotoxicity, and cardiac arrest	Dufton and Hider (1988), Chen et al. (2007), Ponnappa et al. (2008), Dutta et al. (2017)
(iii) Short-chain or long-chain neurotoxin (NTx)		(i) α-Neurotoxins (αNTx) bind to muscle (α1) nicotinic receptors and inhibit the acetylcholine from binding to the receptor. (ii) κ-Neurotoxins (κNTx) explicitly bind to neuronal (α3β4) nicotinic receptors	Neurotoxicity by impaired neuromuscular transmission at various postsynaptic sites in the peripheral and central nervous system	Grant and Chiappinelli (1985), Changeux (1990), Kukhtina et al. (2000), Chung et al. (2002), Kini and Doley (2010), Dutta et al. (2017, 2019)

(iii) Muscarinic toxins (MTLP) target muscarinic acetylcholine receptors (mAChRs)			
(iv) Nonconventional 3FTxs		Reversible and partial reversible binding to muscle (α1) nAChR and neuronal α7 nAChR, respectively	Weak neurotoxicity but the cognate complex of weak neurotoxins/ nonconventional three-finger toxins with PLA$_2$ enhances cytotoxicity on target cells	Nirthanan et al. (2003), Mukherjee (2008a, 2010), Kini and Doley (2010)
Cobra venom factor (CVF)	~150 kDa	Binding of CVF to factor B to form a complex is subsequently cleaved by factor D to form the bimolecular complex CVF, Bb which is a C3/C5 convertase that cleaves both complement components C3 and C5	Complement activation; however, nontoxic in isolation	Kock et al. (2004), Vogel and Fritzinger (2010)
Cysteine-rich secretory protein (CRISP)	20–30	Blocking of ion channels	Pathophysiology of snakebite is unknown	Yamazaki and Morita (2004), Sunagar et al. (2015), Dutta et al. (2017), Patra et al. (2019)
Nerve growth factor (NGF)	12–18 (monomer) 25–54 (dimer)	Molecular mechanism of action in prey or victims is unexplored. Forms cognate complex with PLA$_2$ and cytotoxins from cobra venom, high affinity binding to TrkA receptor of PC-12 and breast cancer (MCf-7 and MDA-MB-231) cells	Toxicity is unknown; however, enhances cytotoxicity of PLA$_2$-cytotoxin cognate complex of cobra venom	Trummal et al. (2011), Kalita et al. (2017), Dutta et al. (2019), Islam et al. (2020)
Ohanin-like protein	~12	Perhaps acts on the central nervous system	Retarding the locomotion and induction of hyperalgesia in mice	Pung et al. (2005), Dutta et al. (2017)
Kunitz-type serine protease inhibitor (KSPI)	6–8	Inhibits plasmin and thrombin, binds to platelet GPIIb/IIIa receptor by RGD-independent manner	Blocking of ion channels, interference with blood coagulation, inflammation, and fibrinolysis. Synergistic effect with other components to enhance the lethality	Mukherjee et al. (2014b), Mukherjee and Mackessy (2014), Mukherjee et al. (2014b), Mukherjee et al. (2016a), Kalita

(continued)

Table 3.3 (continued)

Nonenzymatic toxin family	Molecular mass range (kDa)	Mechanism of action	Pharmacological activity	References
			of venom. Shows concentration-dependent aggregation or disaggregation of platelets	et al. (2017), Thakur and Mukherjee (2017b), Kalita et al. (2019)
C-type lectins (Snaclecs)	8–16 (monomer)	Binding with high affinity to blood coagulation factors IX and/or X, target membrane, and platelet receptors	Diverse biological functions including coagulopathy, platelet modulation	Clemetson (2012), Arlinghaus and Eble (2012); Zhang et al. (2012), Mukherjee et al. (2014c)
Disintegrins (predominated in Viperidae)	4–15	High-affinity binding to extracellular β-subunit of integrin receptors present on platelets and other endothelial cells	Inhibits platelet aggregation and integrin-dependent cell adhesion, cell cytotoxicity	Betzel et al. (2005), Calvete (2005), Saviola et al. (2015, 2016)
Natriuretic peptides (NP)	4–20	Activates membrane-bound guanylyl cyclase receptors	Regulation of pressure-volume homeostasis, sustained and prolonged reduction in blood pressure without renal effects	Pierre et al. (2006), Sridharan and Kini (2015), Dutta et al. (2017)

Their mechanisms of action and pharmacological properties are also shown

3.3   Variation in Snake Venom Composition and Its Impact on the Pathogenesis of Snakebite and Antivenom Treatment

Variations in snake venom composition, a well-known phenomenon, determine the pathophysiological symptoms from snakebite. The great variation in venom composition is due to variation among individual snakes, geographical origins, and age of the snakes (Taborsk & Kornalik, 1985; Meier, 1986; Jayanthi & Gowda, 1988; Daltry et al., 1996; Tsai et al., 1996; Mukherjee & Maity, 1998; Modahl et al., 2016; Kalita et al., 2018c; Kalita & Mukherjee, 2019). Gene mutation, a primary mechanism of evolution, plays an important role in causing variation in venom composition among closely related species or even within the same species of snake (Fry, 2005; Fry et al., 2012; Gibbs & Mackessy, 2009; Richards et al., 2012).

Because of the variation in venom composition, the pathogenesis that develops after snakebite is complex, and the clinical manifestations depend on the qualitative composition and quantitative distribution of different components of venom toxins (Warrell, 1989; Stocker, 1990a, 1990b; Mukherjee & Maity, 1998; Kalita et al., 2018c; Kalita & Mukherjee, 2019; Chanda et al., 2018a; Chanda & Mukherjee, 2020a, 2020b; Mukherjee et al., 2021). For example, Russell's viper venom from southern India differs from that of western and northern India in terms of lethal potencies (Jayanthi & Gowda, 1988). Although *Naja naja* and *Naja kaouthia* are closely related species, they differ in their venom composition (Mukherjee, 1998; Mukherjee & Maity, 2002). The venom of the former is more toxic than that of the latter and the antivenom raised against the venom of *N. naja* is hardly effective in neutralizing the pharmacological and pathological effects of *N. kaouthia* venom of the same geographical origin. Moreover, *N. kaouthia* venom compared to *N. naja* venom demonstrated less immuno-recognition by the polyvalent antivenom (Mukherjee & Maity, 2002; Chanda, Patra, et al., 2018b). Variations in the composition of snake venom are well documented to result in significant deviations in the immunological cross-reactivity between commercial antivenoms and neutralizing potencies of antivenom (Mukherjee & Maity, 2002; Kumar & Gowda, 2006; Harrison et al., 2011; Keyler et al., 2013; Kalita et al., 2018c; Senji Laxme et al., 2019; Chanda & Mukherjee, 2020b). The variations in venom composition must be considered carefully during antivenom production as the antivenom raised against the venom of one population of snakes from a particular location may be less effective against the venom of another population of snakes from another location. This can occur even when both populations belong to the same species (Mukherjee & Maity, 2002; Kumar & Gowda, 2006; Harrison et al., 2011; Keyler et al., 2013; Kalita et al. 2018a,b,c; Senji Laxme et al., 2019).

3.4   Evolution of Genes for the Toxins in Snake Venom

The evolution of snake venom is debatable though several hypotheses have been proposed to explain how it accounts for the vast expansion of snakes around the world (Fry et al., 2012). The link between the origin of venom and the rapid divergence of snakes during the Cenozoic period indicates a shift from a mechanical (constriction) method to a highly specialized venom-mediated (biochemical) method to subdue the prey of advanced snakes including the "Big Four" venomous snakes of India (Fry et al., 2012; Lomonte et al., 2014). The common biological pathways and similar mechanisms for the biological action to exert toxicity into prey by the venom of snakes and other taxa (i.e., spiders and cone snails) unequivocally suggest the convergent evolution of venom (Casewell et al., 2013). Two major hypotheses have been put forward to explain the evolution of venom: (a) the Toxicofera hypothesis and (b) the independent origin hypothesis.

3.4.1   Toxicofera Hypothesis

Traditional phylogenetic trees based on species morphology indicate that around 100 million years ago venom originated in multiple branches of the Squamata (the largest order of reptiles comprised of snakes, lizards, and amphibians) (Fry et al., 2012). Recent studies have shown the occurrence of identical venom proteins in numerous lizards within a proposed clade of scaled reptiles (named Toxicofera) (Fry et al., 2012). This idea suggests that the origin of venom, containing only a few proteins (toxins), occurred at around 170 million years ago. Subsequently, the venoms from different lineages became diversified and developed independently. This included the independent evolution of front-fanged venom delivery from the ancestral rear-fanged venom delivery system, according to their prey specificity (Fry & Wüster, 2004; Fry et al., 2012).

3.4.2   Independent Origin Hypothesis

This hypothesis rejected the Toxicofera hypothesis and proposed that during evolution snake venom progressed independently in a number of families (Hargreaves, Swain, Logan, & Mulley, 2014a). Researchers also suggested that the progression of most families of the venom gene took place frequently in highly venomous caenophidian snakes (Reyes-Velasco et al., 2015).

3.5 Mechanism of the Evolution and Diversification of Venom Proteins

The evolution and diversification of snake venom proteins resulted in the formation of new proteins (toxins) with expanded biological activities. Nevertheless, the genomic and evolutionary origins of most venom toxins have been unclear (Giorgianni et al., 2020). The following subsections describe the theories to explain the evolution of venom toxins.

3.5.1 Accelerated Evolution of Venom Protein Genes

Several hypotheses have been proposed to explain the mechanism of evolution and diversification of snake venoms. One of the earlier ideas was that gene duplications in other tissues and their expression in the venom glands gave rise to diverse venom proteins via natural selection or positive Darwinian selection (Fry, 2005; Casewell et al., 2011, 2013). This hypothesis of gene duplication in tissues other than the salivary glands has been debated. Hargreaves, Swain, Hegarty, et al. (2014b) proposed another hypothesis where the duplication of salivary protein genes was restricted to the venom gland that resulted in the diversification of snake venom proteins. Studies involving the molecular origin and evolution of snake venom proteins, derived from a phylogenetic analysis of toxin sequences and associated body proteins, suggested a correlation between the diversification of tissue types from the toxin recruitment genes and the venom proteins (Fry, 2005). Two types of toxins (CRISP and kallikrein), which have been shown to essentially represent the modified proteins of venom, were previously present in the ancestral salivary tissue (Fry, 2005). Thus, toxin types originating from extensively cysteine cross-linked proteins flourished into newly derived, functionally diverse, toxin multigene families. Interestingly, toxin genes, compared to nontoxin genes in animals, mostly underwent quick accelerated evolution to produce an array of venom proteins to be more effective and competitive in capturing prey, and to augment their predator defense (Ogawa et al., 1995; Casewell et al., 2011; Kini, 2018; Shibata et al., 2019).

The mechanism of accelerated evolution of venom protein genes has been reviewed by Kini (2018). Several theories are involved, such as (i) frequent mutations in exons compared to introns and nonsynonymous substitutions in exons; (ii) high frequency of point mutations; (iii) accelerated segment switch in exons to alter targeting (ASSET); (iv) rapid accumulation of variations in exposed residues (RAVERs); (v) alteration in intron-exon boundary; (vi) deletion of exon; (vii) loss/gain of domains through recombination; and (viii) evolution of these genes through exonization and intronization (Kini, 2018). The proposed mechanism suggests that venom toxin functions are altered through drastic changes in their molecular surface via insertion of new exons and removal of exons. This would empower the toxins to expand their capability to target newer receptors, ion channels, and enzymes for efficient prey capture and defense (Kini, 2018).

3.5.2   Selection Pressure for Rapid Adaptive Evolution

Alternative splicing with gene duplication has also been suggested to be a mechanism of venom diversification (Casewell et al., 2013). The banded krait (*B. fasciatus*) is an example of gene diversification. Gene loss of specific venom components may also be a mechanism for venom diversification (Dowell et al., 2016). Proteomic analyses have shown the occurrence of varying numbers of protein toxins in the venoms of the "Big Four" snakes of the Indian subcontinent (Chanda et al., 2018a,b; Chanda & Mukherjee, 2020a, b; Kalita et al., 2017, 2018a,b,c,d; Dutta et al., 2017; Faisal et al., 2018; Mukherjee et al., 2016b; Oh et al., 2017; Patra et al., 2017, 2019, 2020; Sharma et al., 2015; Sintiprungrat et al., 2016). The protein diversification also leads to geographical variation in venom composition in the same species of snake (Kalita et al., 2018c; Kalita & Mukherjee, 2019; Chanda & Mukherjee, 2020a).

3.5.3   Diet and Snake Venom Evolution

Whether or not diet plays an important role in the evolution of snake venom is a long debatable issue. Daltry et al. (1996) produced convincing evidence that correlated the role of diet or pre-specificity in the evolution of pit viper (*Calloselasma rhodostoma*, Viperidae) venom. Partial Mantel tests and independent contrasts were used to evaluate the probable cause of the geographical variation in venom composition; however, Daltry et al. (1996) overruled the influence of both contemporary gene flow (estimated from geographical proximity) and phylogenetic relationships (assessed by analyzing mitochondrial DNA) on populations with venom evolution (Daltry et al., 1996). The authors concluded that the key function of viperid venom is to immobilize and digest prey and the vulnerability of prey against the venom may vary; therefore, geographical variation in venom composition reveals the natural selection for feeding on local prey (Daltry et al., 1996). In another study, the venom of saw-scaled vipers showed a high degree of variability that depended on its feeding species (Richards et al., 2011). The venoms of arthropod-feeding species of *Echis* demonstrated significantly higher toxicity to normal scorpion prey compared to the species of *Echis* that fed largely on vertebrate prey. Thus, changes in diet appeared to be correlated to snake venom evolution (Richards et al., 2011). These studies suggest that snake venom has adapted to kill its natural, available prey.

3.6   Biological Functions of Venom

Venom has multiple activities, some of which are mentioned below.

3.6.1   Prey-Specific Venom Toxicity

A large number of snake venoms, thought to have evolved through positive Darwinian selection, show acute toxicity only to a certain taxon. This phenomenon is known as prey-specific toxicity and its major purpose is to subdue preferred prey species. Examples include the venom of the mangrove snake (*Boiga dendrophila*) that is specifically toxic to birds (Casewell et al., 2013), and *Echis* venom that shows significantly higher toxicity to natural scorpion prey (Richards et al., 2011). Further, the venoms of four species of *Sistrurus* (pit viper) have demonstrated significant variation in their toxicity to mice, which is correlated to differences in the muscle physiology of the various prey animals (Gibbs & Mackessy, 2009). The phylogenetic analysis of these species of snakes suggests that their venoms evolved to feed on a mammal-based diet (Gibbs & Mackessy, 2009). Daltry et al. (1996) also suggested a pre-specificity in the evolution of pit viper (*Calloselasma rhodostoma*, Viperidae) venom. In conclusion, the prey-specific toxicity helps to ensure a rapid death of the prey. Nevertheless, the venoms of several snakes have potent toxicity against taxa that are not their preferred prey or that are consumed less often (Gibbs & Mackessy, 2009).

3.6.2   Immobilization and Predigestion of Prey

Several toxicologists have suggested that, in addition to killing specific prey, snake venom plays a role in digesting the prey before it is swallowed (Zeller, 1948; Thomas & Pough, 1979). Moreover, they key function of viperid venom is to immobilize and digest prey (Daltry et al., 1996; Richards et al., 2011). Snake venom metalloproteases, in addition to contributing to a swift death of the prey animal, also assist in digesting the prey (Mackessy, 2010). Since most of the venomous snakes feed on fast-moving agile prey, once injected with venom, the bitten prey cannot move ahead and can easily be captured by the snake.

3.6.3   Prey Re-localization

According to this hypothesis, the venom has an adaptive function that allows the snake to track or follow the pathway of a prey animal it has bitten in an environment where the same species of other unbitten animals are present (Saviola et al., 2013). According to the hypothesis, disintegrin proteins (a nonenzymatic component of snake venom) of western diamondback rattlesnakes (*Crotalus atrox*) play a significant role in changing the odor of bitten animals that helps the snake track the prey (Saviola et al., 2013).

3.7   Indian Snake Venom Proteins: A Treasure House of Drug Prototypes and Diagnostic Tool

Toxins from the Elapidae and Viperidae families of Indian snake venoms have great potency, affinity, and selectivity for primarily affecting the hemostasis of the victim or prey, targeting blood coagulation proteins and/or platelets and the neuromuscular and cardiovascular systems (Mukherjee et al., 2000; Mukherjee & Mackessy, 2013; Mukherjee & Maity, 2002; Thakur & Mukherjee, 2017a; Kalita & Mukherjee, 2019; Kalita et al., 2021). Many studies have shown that this spectrum of medically significant toxins need to be considered as candidates for lifesaving drug prototypes (Mukherjee et al., 2011; Koh & Kini, 2012; Thakur & Mukherjee, 2017a, 2017b; Kalita & Mukherjee, 2019). Several of these venom proteins, including those of the "Big Four" venomous snakes of India, are often nontoxic in isolation but can show a high affinity for binding to their targets in very low concentrations. These properties may allow them to be developed as novel drugs to combat several diseases (Mukherjee & Mackessy, 2013, 2014; Mukherjee et al., 2014b,c; Kalita & Mukherjee, 2019). Several toxicologists have unequivocally pointed to the beneficial application of comparatively nontoxic constituents of Indian snake venoms and advocated that such proteins, for example, snake venom anticoagulant proteins, have a high affinity for binding with thrombin and FXa and may serve as drug prototypes in preventing and/or treating thrombosis-associated cardiovascular diseases (Stocker, 1990a, 1990b; Mukherjee et al., 2011; Koh & Kini, 2012; Kalita & Mukherjee, 2019). These venom proteins are represented by phospholipase A_2 (PLA_2), proteases (SVMPs and SVSPs), L-amino acid oxidases (LAAO), Kunitz-type serine protease inhibitors (KSPI), snaclec, anticoagulant, and angiogenic peptides (Kalita & Mukherjee, 2019) (Table 3.3).

In addition to snake venom proteins, some polypeptides in the "Big Four" snake venoms are promising candidates for peptide-based drugs. Mukherjee et al. (2014d) demonstrated that a 3.9 kDa novel polypeptide named RVVAP that had been purified from eastern India (EI) *D. russelii* venom induced angiogenesis in EA. hy926 cells (human umbilical vein cell line) in vitro. The authors suggested that RVVAP could serve as a cicatrization drug prototype and be used in wound healing (Thakur et al., 2019). Another 4.4 kDa polypeptide purified from EI RVV (named Ruviprase) demonstrated anticoagulant activity via dual inhibition of thrombin and factor Xa, the two key components in the blood coagulation cascade (Thakur et al., 2014). A dual inhibitor of thrombin and FXa is not yet available commercially. When injected into experimental mice at a dose of 2.0 mg/kg, Ruviprase did not show acute toxicity but demonstrated in vivo anticoagulant activity, indicating that this peptide would be an appropriate candidate as an anticoagulant drug prototype to prevent blood coagulation. Remarkably, besides showing anticoagulant activity, Ruviprase also exhibits antiproliferative activity against MCF-7 cells and induces apoptosis through an intrinsic pathway of apoptosis (Thakur et al., 2016). Thus, Ruviprase is a very promising drug prototype for the treatment of breast cancer.

The pharmacologically active drug prototype toxins, which have been purified and characterized, especially those from the Indian Big Four snake venoms, are listed in Table 3.4.

Table 3.4 Some promising drug prototypes characterized from the "Big Four" Indian snake venoms

Name of drug prototype/toxin	Purified from	Toxin class	Biological activity	Molecular mass	References
A. Antiplatelet, antithrombotic drug					
EC-I-PLA$_2$	E. carinatus	PLA$_2$	Antiplatelet	16.0	Kemparaju et al. (1994)
EC-IV-PLA$_2$	E. carinatus	PLA$_2$	Antiplatelet	14.0	Kemparaju et al. (1999)
NN-PF3	N. naja	SVMP	Anticoagulant and antiplatelet	68.0	Jagadeesha et al. (2002)
Daboialectin	D. russelii		Inhibition of platelet aggregation	18.5	Pathan et al. (2017)
B. Blood thinner, antithrombotic drug					
RVVA-PLA2-I	D. russelii	PLA$_2$	Catalytic, anticoagulant, and membrane-damaging activity	28.5	Saikia et al. (2011, 2012)
NEUPHOLIPASE	D. russelii	PLA$_2$	Catalytic and anticoagulant activity	13.0	Saikia et al. (2013)
RVVPLA$_2$	D. russelii	PLA$_2$	Catalytic and anticoagulant activity	13.8	Mukherjee (2014a)
RVsnaclec	D. russelii	Snaclec	Anticoagulant	66.3	Mukherjee et al. (2014c)
Nk-PLA$_2$α	N. kaouthia	PLA$_2$	Catalytic and anticoagulant activity; inhibition of thrombin and/or FXa	13.4	Mukherjee et al. (2014a)
Nk-PLA$_2$β	N. kaouthia	PLA$_2$	Catalytic and anticoagulant activity; inhibition of thrombin and/or FXa	13.2	Mukherjee et al. (2014a)
Rusvikunin	D. russelii	KSPI	Inhibition of thrombin and/or FXa	6.9	Mukherjee et al. (2014b)
Rusvikunin II	D. russelii	KSPI	Inhibition of thrombin and/or FXa	7.1	Mukherjee and Mackessy (2014)
NnPLA$_2$-I	N. naja	Phospholipase A$_2$	Catalytic, anticoagulant, and antiplatelet activity; inhibition of thrombin	15.2	Dutta et al. (2015)
Daboxin P	D. russelii	Phospholipase A$_2$	Catalytic and anticoagulant activity	13.6	Sharma et al. (2016)

(continued)

Table 3.4 (continued)

Name of drug prototype/toxin	Purified from	Toxin class	Biological activity	Molecular mass	References
Ruviprase	*D. russelii*	Polypeptide	Dual inhibition of thrombin and FXa	4.4	Thakur et al. (2014)
C. Blood clot-promoting drug					
RVBCMP	*D. russelii*	SVMP	Fibrin(ogeno)lytic and procoagulant	15.0	Mukherjee (2008b)
Rusviprotease	*D. russelii*	SVMP	Fibrin(ogeno)lytic and procoagulant activity	26.8	Thakur et al. (2015)
D. Treatment of hyperfibrinogenemia-associated and other cardiovascular disorders					
Russelobin	*D. russelii*	SVSP	Fibrin(ogeno)lytic, procoagulant, and thrombin-like activity	51.3	Mukherjee and Mackessy (2013)
RV-FVPα	*D. russelii*	SVSP	Fibrinogenolytic, esterolytic, and procoagulant activity	32.9	Mukherjee (2014b)
RV-FVPβ	*D. russelii*	SVSP	Fibrinogenolytic, esterolytic, and procoagulant activity	33.3	Mukherjee (2014b)
RV-FVPγ	*D. russelii*	SVSP	Fibrinogenolytic, esterolytic, and procoagulant activity	33.3	Mukherjee (2014b)
RV-FVPδ	*D. russelii*	SVSP	Fibrinogenolytic, esterolytic, and procoagulant activity	34.5	Mukherjee (2014b)
Natriuretic peptides	Eastern India *N. naja*	Nonenzymatic proteins	NP analogs lower blood pressure and circulating volume	12.9	Dutta et al. (2017); Sridharan et al. (2020)
E. Anticancer drug					
drCT-I	*D. russelii*	Heat-stable nonenzymatic protein	Antiproliferative activity	7.2	Gomes et al. (2007)
drCT-II	*D. russelii*	Uncharacterized protein	Anticancer activity	6.2	Gomes et al. (2015)
NN-32	*N. naja*	Nonenzymatic protein	Antiproliferative activity, cytotoxicity on EAC cells	6.7	Das et al. (2011); Gomes et al. (2014)

Rusvinoxidase	*D. russelii*	LAAO	Catalytic activity and cytotoxicity against MCF-7 breast cancer cell	57.0	Mukherjee et al. (2015, 2018)
Ruviprase	*D. russelii*	Nonenzymatic polypeptide	Cytotoxicity against MCF-7 breast cancer cell	4.4	Thakur et al. (2016)
F. Cicatrization drug and wound healing					
RVVAP	*D. russelii*	Nonenzymatic polypeptide	Angiogenesis	3.9	Mukherjee et al. (2014d), Thakur et al. (2019)
G. Anticonvulsant drug					
KC-MMTx	*Ophiophagus hannah*	Unsaturated aliphatic acid	Protection against amphetamine aggregate toxicity in mice	0.28	Saha et al. (2006)
H. Antibacterial peptide					
Antibacterial peptide	*N. naja*	Nonenzymatic peptide	Potent antibacterial activity against Gram-negative bacteria (*Escherichia coli, Pseudomonas aeruginosa,* and *Vibrio cholera*) and Gram-positive bacteria (*Staphylococcus aureus, Enterococcus faecalis, Streptococcus pneumoniae, Streptococcus pyogenes,* and *Bacillus subtilis*)	2.4	Sachidananda et al. (2007)

Further preclinical studies on the toxicity, pharmacokinetics, and pharmacodynamics of purified venom components (drug prototypes) and/or their derivatives are necessary for the development and production of Indian snake venom-based therapeutics.

References

Ahmed, M., Rocha, J. B., Morsch, V. M., & Schetinger, M. R. (2009). Snake venom acetylcholinesterase. In S. Mackessy (Ed.), *Handbook of venoms and toxins of reptiles* (pp. 207–219). CRC Press.

Ande, S. R., Kommoju, P. R., Draxl, S., Murkovic, M., Macheroux, P., Ghisla, S., & Ferrando-May, E. (2006). Mechanisms of cell death induction by L-amino acid oxidase, a major component of ophidian venom. *Apoptosis, 11*, 1439–1451.

Arlinghaus, F. T., & Eble, J. A. (2012). C-type lectin-like proteins from snake venoms. *Toxicon, 60*, 512–519.

Betzel, C., Sharma, S., Singh, T. P., Perbandt, M., Yadav, S., Kaur, P., & Bilgrami, S. (2005). Crystal structure of the disintegrin heterodimer from saw-scaled viper (Echis carinatus) at 1.9 A resolution. *Biochemistry, 44*(33), 11058–11066.

Bjarnason, J. B., & Fox, J. W. (1994). Hemorrhagic metalloproteinases from snake venoms. *Pharmacology & Therapeutics, 62*, 325–372.

Bilwes, A., Rees, B., Moras, D., Menez, R., & Menez, A. (1994). X-ray structure at 1.55 Å of toxin γ, a cardiotoxin from *Naja nigricollis* venom: Crystal packing reveals a model for insertion into membranes. *Journal of Molecular Biology, 239*, 122–136.

Calvete, J. (2005). Structure-function correlations of snake venom disintegrins. *Current Pharmaceutical Design, 11*(7), 825–835.

Casewell, N. R., Wagstaff, S. C., Harrison, R. A., Renjifo, C., & Wuster, W. (2011). Domain loss facilitates accelerated evolution and neofunctionalization of duplicate snake venom metalloproteinase toxin genes. *Molecular Biology and Evolution, 28*, 2637–2649.

Casewell, N. R., Wuster, W., Vonk, F. J., Harrison, R. A., & Fry, B. G. (2013). Complex cocktails: the evolutionary novelty of venoms. *Trends in Ecology & Evolution, 28*(4), 219–229.

Chanda, A., Kalita, B., Patra, A., Senevirathne, W. D., & Mukherjee, A. K. (2018a). Proteomic analysis and antivenomics study of Western India *Naja naja* venom: Correlation between venom composition and clinical manifestations of cobra bite in this region. *Expert Review of Proteomics, 16*(2), 171–184.

Chanda, A., Patra, A., Kalita, B., & Mukherjee, A. K. (2018b). Proteomics analysis to compare the venom composition between *Naja naja* and *Naja kaouthia* from the same geographical location of eastern India: correlation with pathophysiology of envenomation and immunological cross-reactivity towards commercial polyantivenom. *Expert Review of Proteomics, 15*(11), 949–961.

Chanda, A., & Mukherjee, A. K. (2020a). Mass spectrometry analysis to unravel the venom proteome composition of Indian snakes: Opening new avenues in clinical research. *Expert Review of Proteomics, 17*, 411–423.

Chanda, A., & Mukherjee, A. K. (2020b). Quantitative proteomics to reveal the composition of Southern India spectacled cobra (Naja naja) venom and its immunological cross-reactivity towards commercial antivenom. *International Journal of Biological Macromolecules, 160*, 224–232.

Changeux, J. P. (1990). The TiPS lecture. The nicotinic acetylcholine receptor: An allosteric protein prototype of ligand-gated ion channels. *Trends in Pharmacological Sciences, 11*, 485–492.

Chen, K.-C., Kao, P.-H., Lin, S.-R., & Chang, L.-S. (2007). The mechanism of cytotoxicity by *Naja naja* atra cardiotoxin 3 is physically distant from its membrane-damaging effect. *Toxicon, 50*, 816–824.

Chen, L.-W., Kao, P.-H., Fu, Y.-S., Lin, S.-R., & Chang, L.-S. (2011). Membrane-damaging activity of Taiwan cobra cardiotoxin 3 is responsible for its bactericidal activity. *Toxicon, 58*, 46–53.

Chen, H.-S., Tsai, H.-Y., Wang, Y.-M., & Tsai, I.-H. (2008). P-III hemorrhagic metalloproteinases from Russell's Viper venom: Cloning, characterization, phylogenetic and functional site analyses. *Biochimie, 90*, 1486–1498.

Choumet, V., Saliou, B., Fideler, L., Chen, Y. C., Gubensek, F., Bon, C., & Delot, E. (1993). Snake-venom phospholipase A_2 neurotoxins. Potentiation of a single-chain neurotoxin by the chaperon subunit of a two-component neurotoxin. *European Journal of Biochemistry, 211*, 57–62.

Chung, C., Wu, B. N., Yang, C. C., & Chang, L. S. (2002). Muscarinic toxin-like proteins from Taiwan banded krait (*Bungarus multicinctus*) venom: Purification, characterization and gene organization. *Biological Chemistry, 383*, 1397–1406.

Clemetson, K. J. (2012). Snaclecs (snake C-type lectins) that inhibit or activate platelets by binding to receptors. *Toxicon, 56*, 1236–1246.

Cousin, X., Creminon, C., Grassi, J., Meflah, K., Cornu, G., Saliou, B., Bon, S., Massoulie, J., & Bon, C. (1996). Acetylcholinesterase from *Bungarus* venom: A monomeric species. *FEBS Letters, 387*, 196–200.

Daltry, J. C., Wuster, W., & Thorpe, R. S. (1996). Diet and snake venom evolution. *Nature, 379*, 537–540.

Das, T., Bhattacharya, S., Halder, B., Biswas, A., Das Gupta, S., Gomes, A., & Gomes, A. (2011). Cytotoxic and antioxidant property of a purified fraction (NN-32) of Indian *Naja naja* venom on Ehrlich ascites carcinoma in BALB/c mice. *Toxicon, 57*, 1065–1072.

Debnath, A., Saha, A., Gomes, A., Biswas, S., Chakrabarti, P., Giri, B., Biswas, A. K., Dasgupta, S., & Gomes, A. (2010). A lethal cardiotoxic–cytotoxic protein from the Indian monocellate cobra (*Naja kaouthia*) venom. *Toxicon, 56*, 569–579.

Deufela, A., & Cundall, D. (2006). Functional plasticity of the venom delivery system in snakes with a focus on the poststrike prey release behavior. *Zoologischer Anzeiger – A Journal of Comparative Zoology, 245*(3–4), 249–267.

Dhananjaya, B. L., & D'Souza, C. J. (2010). The pharmacological role of nucleotidases in snake venoms. *Cell Biochemistry and Function, 28*, 171–177.

Doley, R., & Mukherjee, A. K. (2003). Purification and characterization of an anticoagulant phospholipase A_2 from Indian monocled cobra (*Naja kaouthia*) venom. *Toxicon, 41*, 81–91.

Doley, R., King, G. F., & Mukherje, A. K. (2004). Differential hydrolysis of erythrocyte and mitochondrial membrane phospholipids by two phospholipase A_2 isoenzymes (NK-PLA$_2$-I and NK-PLA$_2$-II), from Indian monocled cobra *Naja kaouthia* venom. *Archives of Biochemistry and Biophysics, 425*, 1–13.

Dowell, N. L., Giorgianni, M. W., Kassner, V. A., Selegue, J. E., Sanchez, E. E., & Carroll, S. B. (2016). The deep origin and recent loss of venom toxin genes in rattlesnakes. *Current Biology, 26*(18), 2434–2445.

Du, X.-P., & Clemetson, K. J. (2002). Snake venom l-amino acid oxidases. *Toxicon, 40*, 659–665.

Dufton, M. J., & Hider, R. C. (1988). Structure and pharmacology of elapid cytotoxins. *Pharmacology & Therapeutics, 36*, 1–40.

Dutta, S., Gogoi, D., & Mukherjee, A. K. (2015). Anticoagulant mechanism and platelet deaggregation property of a non-cytotoxic, acidic phospholipase A_2 purified from Indian cobra (*Naja naja*) venom: Inhibition of anticoagulant activity by low molecular weight heparin. *Biochimie, 110*, 93–106.

Dutta, S., Chanda, A., Islam, T., Patra, A., & Mukherjee, A. K. (2017). Proteomic analysis to unravel the complex venom proteome of eastern India *Naja naja*: Correlation of venom composition with its biochemical and pharmacological properties. *Journal of Proteomics, 156*, 29–39.

Dutta, S., Archana Sinha, A., Dasgupta, S., & Mukherjee, A. K. (2019). Binding of a *Naja naja* venom acidic phospholipase A_2 cognate complex to membrane-bound vimentin of rat L6 cells:

Implications in cobra venom-induced cytotoxicity. *Biochimica et Biophysica Acta, Biomembranes, 1861*, 958–977.

Elliott, W. B. (1978). Chemistry and immunology of reptilian venoms. In C. Gans (Ed.), *Biology of the Reptilia* (Vol. 8, p. 163). Academic Press.

Faisal, T., Tan, K. Y., Sim, S. M., Quraishi, N., Tan, N. H., & Tan, C. H. (2018). Proteomics, functional characterization and antivenom neutralization of the venom of Pakistani Russell's viper (*Daboia russelii*) from the wild. *Journal of Proteomics, 183*, 1–13.

Fletcher, J. E., & Jiang, M. S. (1993). Possible mechanisms of action of cobra snake venom cardiotoxins and bee venom melittin. *Toxicon, 31*, 669–695.

Fox, J. W. (2013). A brief review of the scientific history of several lesser-known snake venom proteins: L-amino acid oxidases, hyaluronidases and phosphodiesterases. *Toxicon, 62*, 75–82.

Frobert, Y., Créminon, C., Cousin, X., Rémy, M. H., Chatel, J. M., Bon, S., & Grassi, J. (1997). Acetylcholinesterases from Elapidae snake venoms: biochemical, immunological and enzymatic characterization. *Biochimica et Biophysica Acta (BBA) - Protein Structure and Molecular Enzymology, 1339*, 253–267.

Fry, B. G. (2005). From genome to venome: Molecular origin and evolution of the snake venom proteome inferred from phylogenetic analysis of toxin sequences and related body proteins. *Genome Research, 15*(3), 403–420.

Fry, B. G., & Wüster, W. (2004). Assembling an Arsenal: origin and evolution of the snake venom proteome inferred from phylogenetic analysis of toxin sequences. *Molecular Biology and Evolution, 21*, 870–883.

Fry, B. G., Casewell, N. R., Wuster, W., Vidal, N., Young, B., & Jackson, T. N. W. (2012). The structural and functional diversification of the Toxicofera reptile venom system. *Toxicon, 60*(4), 434–448.

Gibbs, H. L., & Mackessy, S. P. (2009). Functional basis of a molecular adaptation: Prey-specific toxic effects of venom from *Sistrurus* rattlesnakes. *Toxicon, 53*(6), 672–679.

Grant, G. A., & Chiappinelli, V. A. (1985). kappa-Bungarotoxin: complete amino acid sequence of a neuronal nicotinic receptor probe. *Biochemistry, 24*, 1532–1537.

Girish, K. S., Mohanakumari, H. P., Nagaraju, S., Vishwanath, B. S., & Kemparaju, K. (2004a). Hyaluronidase and protease activities from Indian snake venoms: neutralization by *Mimosa pudica* root extract. *Fitoterapia, 75*, 378–380.

Girish, K. S., Shashidharamurthy, R., Nagaraju, S., Gowda, T. V., & Kemparaju, K. (2004b). Isolation and characterization of hyaluronidase a "spreading factor" from Indian cobra (*Naja naja*) venom. *Biochimie, 86*(3), 193–202.

Giorgianni, M. W., Dowell, N. L., Griffin, S., Kassner, V. A., Selegue, J. E., & Carroll, S. B. (2020). The origin and diversification of a novel protein family in venomous snakes. *Proceedings of the National Academy of Sciences, 117*(20), 10911–10920.

Gomes, A., Choudhury, S. R., Saha, A., Mishra, R., Giri, B., Biswas, A. K., Debnath, A., & Gomes, A. (2007). A heat stable protein toxin (drCT-I) from the Indian Viper (*Daboia russelii russelii*) venom having antiproliferative, cytotoxic and apoptotic activities. *Toxicon, 49*, 46–56.

Gomes, A., Datta, P., Das, T., Biswas, A. K., & Gomes, A. (2014). Anti arthritic and anti-inflammatory activity of a cytotoxic protein NN-32 from Indian spectacle cobra (*Naja naja*) venom in male albino rats. *Toxicon, 90*, 106–110.

Gomes, A., Biswas, A. K., Bhowmik, T., Saha, P. P., & Gomes, A. (2015). Russell's Viper venom purified toxin Drct-II inhibits the cell proliferation and induces G1 cell cycle arrest. *Translational Medicine, 5*. TM (open access).

Guieu, R., Rosso, J. P., & Rochat, H. (1994). Anticholinesterases and experimental envenomation by *Naja*. *Comparative Biochemistry and Physiology. Part C, Pharmacology, Toxicology & Endocrinology, 109*, 265–268.

Gutiérrez, J. M., Rucavado, A., & Escalante, T. (2009). Snake venom metalloproteinases. Biological roles and participation in the pathophysiology of envenomation. In S. P. Mackessy (Ed.), *Handbook of venom and reptiles* (pp. 115–138). CRC Press.

Hargreaves, A. D., Swain, M. T., Logan, D. W., & Mulley, J. F. (2014a). Testing the Toxicofera: Comparative transcriptomics casts doubt on the single, early evolution of the reptile venom system. *Toxicon, 92*, 140–156.

Hargreaves, A. D., Swain, M. T., Hegarty, M. J., Logan, D. W., & Mulley, J. F. (2014b). Restriction and recruitment—gene duplication and the origin and evolution of snake venom toxins. *Genome Biology and Evolution, 6*(8), 2088–2095.

Harrison, R. A., Cook, D. A., Renjifo, C., Casewell, N. R., Currie, R. B., & Wagstaff, S. C. (2011). Research strategies to improve snakebite treatment: Challenges and progress. *Journal of Proteomics, 74*, 1768–1780.

Islam, T., Majumdar, M., Bidkar, A., Ghosh, S. S., Mukhopadhyay, R., Utkin, Y., & Mukherjee, A. K. (2020). Nerve growth factor from Indian Russell's viper venom (RVV-NGFa) shows high affinity binding to TrkA receptor expressed in breast cancer cells: Application of fluorescence labeled RVV-NGFa in the clinical diagnosis of breast cancer. *Biochimie, 176*, 311–344.

Jackson, K. (2003). The evolution of venom-delivery systems in snakes. *Zoological Journal of the Linnean Society, 137*, 337–354.

Jagadeesha, D. K., Shashidharamurthy, R., Girish, K. S., & Kemparaju, K. (2002). A non-toxic anticoagulant metalloprotease: purification and characterization from Indian cobra (*Naja naja naja*) venom. *Toxicon, 40*, 667–675.

Jayanthi, G. P., & Gowda, T. V. (1988). Geographical variation in India in the composition and potency of Russell's viper (*Vipera russelii*) venom. *Toxicon, 26*, 257–264.

Kalita, B., & Mukherjee, A. K. (2019). Recent advances in snake venom proteomics research in India: a new horizon to decipher the geographical variation in venom proteome composition and exploration of candidate drug prototypes. *Journal of Proteins and Proteomics, 10*, 149–164.

Kalita, B., Patra, A., & Mukherjee, A. K. (2017). Unravelling the proteome composition and immuno-profiling of western India Russell's Viper venom for in-depth understanding of its pharmacological properties, clinical manifestations, and effective antivenom treatment. *Journal of Proteome Research, 16*, 583–598.

Kalita, B., Patra, A., Das, A., & Mukherjee, A. K. (2018a). Proteomic analysis and immuno-profiling of eastern India Russell's viper (*Daboia russelii*) venom: Correlation between RVV composition and clinical manifestations post RV bite. *Journal of Proteome Research, 17*, 2819–2833.

Kalita, B., Patra, A., & Mukherjee, A. K. (2018b). First report of the characterization of a snake venom apyrase (Ruviapyrase) from Indian Russell's viper (*Daboia russelii*) venom. *International Journal of Biological Macromolecules, 111*, 639–648.

Kalita, B., Mackessy, S. P., & Mukherjee, A. K. (2018c). Proteomic analysis reveals geographic variation in venom composition of Russell's Viper in the Indian subcontinent: Implications for clinical manifestations post-envenomation and antivenom treatment. *Expert Review of Proteomics, 15*, 837–849.

Kalita, B., Singh, S., Patra, A., & Mukherjee, A. K. (2018d). Quantitative proteomic analysis and antivenom study revealing that neurotoxic phospholipase A_2 enzymes, the major toxin class of Russell's viper venom from southern India, shows the least immuno-recognition and neutralization by commercial polyvalent antivenom. *International Journal of Biological Macromolecules, 118*, 375–385.

Kalita, B., Dutta, S., & Mukherjee, A. K. (2019). RGD-independent binding of Russell's viper venom Kunitz-type protease inhibitors to platelet GPIIb/IIIa receptor. *Scientific Reports, 9*, 8316.

Kalita, B., Saviola, A. J., & Mukherjee, A. K. (2021). From venom to drugs: A review and critical analysis of Indian snake venom toxins envisaged as anti-cancer drug prototypes. *Drug Discovery Today, 26*(4), 993–1005.

Kang, T. S., Georgieva, D., Genov, N., Murakami, M. T., Sinha, M., Kumar, R. P., Kaur, P., Kumar, S., Dey, S., Sharma, S., Vrielink, A., Betzel, C., Takeda, S., Arni, R. K., Singh, T. P., & Kini, R. M. (2011). Enzymatic toxins from snake venom: structural characterization and mechanism of catalysis. *The FEBS Journal, 278*, 4544–4576.

Kardong, K. V. (1982). The evolution of the venom apparatus in snakes from colubrids to viperids and elapids. *Memórias do Instituto Butantan, 46*, 106–118.

Kardong, K., Weinstein, S., & Smith, T. (2009). Reptile venom glands: Form, function, and future. In S. P. Mackessy (Ed.), *Handbook of venom and reptiles* (pp. 66–91). CRC Press.

Kemparaju, K., Prasad, B. N., & Gowda, V. T. (1994). Purification of a basic phospholipase A_2 from Indian saw-scaled viper (*Echis carinatus*) venom: characterization of antigenic, catalytic and pharmacological properties. *Toxicon, 32*, 1187–1196.

Kemparaju, K., Krishnakanth, T. P., & Gowda, T. V. (1999). Purification and characterization of a platelet aggregation inhibitor acidic phospholipase A_2 from Indian saw-scaled viper (*Echis carinatus*) venom. *Toxicon, 37*, 1659–1671.

Keyler, D. E., Gawarammana, I., Gutiérrez, J. M., Sellahewa, K. H., McWhorter, K., & Malleappa, H. R. (2013). Antivenom for snakebite envenoming in Sri Lanka: The need for geographically specific antivenom and improved efficacy. *Toxicon, 69*, 90–97.

Kini, R. M. (2018). Accelerated evolution of toxin genes: Exonization and intronization in snake venom disintegrin/metalloprotease genes. *Toxicon, 148*, 16–25.

Kini, R. M., & Doley, R. (2010). Structure, function and evolution of three-finger toxins: Mini proteins with multiple targets. *Toxicon, 56*, 855–867.

Kini, R. M., & Gowda, T. V. (1982). Studies on snake venom enzymes: part II—partial characterization of ATPases from Russell's viper (*Vipera russelii*) venom and their interaction with potassium gymnemate. *Indian Journal of Biochemistry & Biophysics, 19*, 342–346.

Kochva, E. (1978). Oral glands of the Reptilia. In C. Gans & K. A. Gans (Eds.), *Biology of the Reptilia* (Vol. 1, pp. 43–161). New York.

Kock, M. A., Hew, B. E., Bammert, H., Fritzinger, D. C., & Vogel, C. W. (2004). Structure and function of recombinant cobra venom factor. *The Journal of Biological Chemistry, 279*, 30836–30843.

Kochva, E. (1987). The origin of snakes and evolution of the venom apparatus. *Toxicon, 25*, 65–106.

Koh, C. Y., & Kini, R. M. (2012). From snake venom toxins to therapeutics-cardiovascular examples. *Toxicon, 59*, 497–506.

Kukhtina, V. V., Weise, C., Muranova, T. A., Starkov, V. G., Franke, P., Hucho, F., & Utkin, Y. N. (2000). Muscarinic toxin-like proteins from cobra venom. *European Journal of Biochemistry, 267*, 6784–6789.

Kumar, A. V., & Gowda, T. V. (2006). Novel non-enzymatic toxic peptide of *Daboia russelii* (Eastern region) venom renders commercial polyvalent antivenom ineffective. *Toxicon, 47*(4), 398–408.

Li, Z. Y., Yu, T. F., & Lian, E. C. (1994). Purification and characterization of L-amino acid oxidase from king cobra (*Ophiophagus hannah*) venom and its effects on human platelet aggregation. *Toxicon, 32*, 1349–1358.

Lomonte, B., Fernandez, J., Sanz, L., Angulo, Y., Sasa, M., Gutierrez, J. M., & Calvete, J. J. (2014). Venomous snakes of Costa Rica: Biological and medical implications of their venom proteomic profiles analyzed through the strategy of snake venomics. *Journal of Proteomics, 105*, 323–339.

Mackessy, S. P. (1991). Morphology and ultrastructure of venom glands of the northern Pacific rattlesnake, *Crotalus viridis oreganus*. *Journal of Morphology, 208*, 109–128.

Mackessy, S. P. (2010). Evolutionary trends in venom composition in the Western Rattlesnakes (*Crotalus viridis* sensu lato): Toxicity vs. tenderizers. *Toxicon, 55*(8), 1463–1474.

Mackessy, S. P., & Baxter, L. M. (2006). Bioweapons synthesis and storage: The venom gland of front-fanged snakes. *Zoologischer Anzeiger, 25*, 147–159.

Mackessy, S. P., & Saviola, A. J. (2016). Understanding biological roles of venoms among the Caenophidia: The importance of rear-fanged snakes. *Integrative and Comparative Biology, 56* (5), 1004–1021.

McCleary, R. J. R., & Kini, R. M. (2013). Non-enzymatic proteins from snake venoms: A gold mine of pharmacological tools and drug leads. *Toxicon, 62*, 56–54.

Meier, J. (1986). Individual and age-dependent variation in the venom of the fer-de-lance (*Bothrops atrox*). *Toxicon, 24*, 41–46.

Meier, J., & Stocker, K. F. (1995). Biology and distribution of venomous snakes of medical importance and the composition of snake venoms. In J. White (Ed.), *Handbook of clinical toxicology of animal venoms and poisons* (pp. 367–412). CRC Press.

Méndez, I., Gutiérrez, J. M., Angulo, Y., Calvete, J. J., & Lomonte, B. (2011). Comparative study of the cytolytic activity of snake venoms from African spitting cobras (*Naja* spp., Elapidae) and its neutralization by a polyspecific antivenom. *Toxicon, 58*, 558–564.

Minton, S. A. (1970). Snake venoms and envenomation. *Clinical Toxicology, 3*(3), 343–345.

Modahl, C. M., Mukherjee, A. K., & Mackessy, S. P. (2016). An analysis of venom ontogeny and prey-specific toxicity in the Monocled Cobra (*Naja kaouthia*). *Toxicon, 119*, 8–20.

Mukherjee, A. K. (1998). *In: Some biochemical properties of cobra and Russell's viper venom and their some biological effects on albino rats.* Burdwan University, Burdwan.

Mukherjee, A. K. (2007). Correlation between the phospholipids domains of the target cell membrane and the extent of *Naja kaouthia* PLA_2 -induced membrane damage: Evidence of distinct catalytic and cytotoxic sites in PLA_2 molecules. *Biochemica et Biophysica Acta, 1770*, 187–195.

Mukherjee, A. K. (2008a). Phospholipase A_2-interacting weak neurotoxins from venom of mono-cled cobra *Naja kaouthia* display cell specific cytotoxicity. *Toxicon, 51*, 1538–1543.

Mukherjee, A. K. (2008b). Characterization of a novel pro-coagulant metalloproteinase (RVBCMP) possessing alpha-fibrinogenase and tissue haemorrhagic activity from venom of *Daboia russelii russelii* (Russell's Viper): Evidence of distinct coagulant and haemorrhagic sites in RVBCMP. *Toxicon, 51*(5), 923–933.

Mukherjee, A. K. (2010). Non-covalent interaction of phospholipase A_2 (PLA_2) and kaouthiotoxin (KTX) from venom of *Naja kaouthia* exhibits marked synergism to potentiate their cytotoxicity on target cells. *Journal of Venom Research, 1*, 37–42.

Mukherjee, A. K. (2013). An updated inventory on properties, pathophysiology and therapeutic potential of snake venom thrombin-like proteases. In S. Chakraborti & N. S. Dhalla (Eds.), *Proteases in health and disease-advances in biochemistry in health and disease* (Vol. 7, pp. 163–180. (ISBN 978-1-4614-9233-7). Springer.

Mukherjee, A. K. (2014a). A major phospholipase A_2 from *Daboia russelii russelii* venom shows potent anticoagulant action via thrombin inhibition and binding with plasma phospholipids. *Biochimie, 99*, 153–161.

Mukherjee, A. K. (2014b). The pro-coagulant fibrinogenolytic serine protease isoenzymes from *Daboia russelii russelii* venom coagulate the blood through factor V activation: Role of glycosylation on enzymatic activity. *PLoS One, 9*(2), e86823. https://doi.org/10.1371/journal.pone.0086823

Mukherjee, A. K., & Maity, C. R. (1998). Composition of *Naja naja* venom sample from three district of West Bengal, Eastern India. *Comparative Biochemistry and Physiology – Part A, 119*, 621–627.

Mukherjee, A. K., & Maity, C. R. (2002). Biochemical composition, lethality and pathophysiology of venom from two cobras--*Naja naja* and *N. kaouthia*. *Comparative Biochemistry and Physiology Part B, Biochemistry & Molecular Biology, 131*, 125–132.

Mukherjee, A. K., & Mackessy, S. P. (2013). Biochemical and pharmacological properties of a new thrombin-like serine protease (Russelobin) from the venom of Russell's viper (*Daboia russelii russelii*) and assessment of its therapeutic potential. *Biochimica et Biophysica Acta, 1830*(6), 3476–3488.

Mukherjee, A. K., & Mackessy, S. P. (2014). Pharmacological properties and pathophysiological significance of a Kunitz-type protease inhibitor (Rusvikunin-II) and its protein complex (Rusvikunin complex) purified from *Daboia russelii russelii* venom. *Toxicon, 89*, 55–66.

Mukherjee, A. K., Ghosal, S. K., & Maity, C. R. (2000). Some biochemical properties of Russell's viper (*Daboia russelii*) venom from eastern India: correlation with clinico-pathological manifestation in Russell's viper bite. *Toxicon, 38*, 163–175.

Mukherjee, A. K., Saikia, D., & Thakur, R. (2011). Medical and diagnostic application of snake venom proteomes. *Journal of Proteins and Proteomics, 2*(1), 31–40.

Mukherjee, A. K., Kalita, B., & Thakur, R. (2014a). Two acidic, anticoagulant PLA$_2$ isoenzymes purified from the venom of monocled cobra *Naja kaouthia* exhibit different potency to inhibit thrombin and factor Xa via phospholipids independent, non-enzymatic mechanism. *PLoS One, 9*(8), e101334.

Mukherjee, A. K., Mackessy, S. P., & Dutta, S. (2014b). Characterization of a Kunitz-type protease inhibitor peptide (Rusvikunin) purified from *Daboia russelii russelii* venom. *International Journal of Biological Macromolecules, 67*, 154–162.

Mukherjee, A. K., Dutta, S., & Mackessy, S. P. (2014c). A new C-type lectin (RVsnaclec) purified from venom of *Daboia russelii russelii* shows anticoagulant activity via inhibition of FXa and concentration-dependent differential response to platelets in a Ca^{2+}-independent manner. *Thrombosis Research, 134*, 1150–1156.

Mukherjee, A. K., Chatterjee, S., Majumdar, S., Saikia, D., Thakur, R., & Chatterjee, A. (2014d). Characterization of a pro-angiogenic, novel peptide from Russell's viper (*Daboia russelii russelii*) venom. *Toxicon, 77*, 26–31.

Mukherjee, A. K., Saviola, A. J., Burns, P. D., & Mackessy, S. P. (2015). Apoptosis induction in human breast cancer (MCF-7) cells by a novel venom L-amino acid oxidase (Rusvinoxidase) is independent of its enzymatic activity and is accompanied by caspase-7 activation and reactive oxygen species production. *Apoptosis, 20*, 1358–1372.

Mukherjee, A. K., Dutta, S., Kalita, B., Jha, D. K., Deb, P., & Mackessy, S. P. (2016a). Structural and functional characterization of complex formation between two Kunitz-type serine protease inhibitors from Russell's viper venom. *Biochimie, 128*, 138–147.

Mukherjee, A. K., Kalita, B., & Mackessy, S. P. (2016b). A proteomic analysis of Pakistan *Daboia russelii russelii* venom and assessment of potency of Indian polyvalent and monovalent antivenom. *Journal of Proteomics, 144*, 73–86.

Mukherjee, A. K., Saviola, A. J., & Mackessy, S. P. (2018). Cellular mechanism of resistance of human colorectal adenocarcinoma cells against apoptosis-induction by Russell's viper venom L-amino acid oxidase (Rusvinoxidase). *Biochimie, 150*, 8–15.

Mukherjee, A. K., Kalita, B., Dutta, S., Patra, A., Maity, C. R., & Punde, D. (2021). Snake envenomation: Therapy and challenges in India. In S. P. Mackessy (Ed.), *Section V: Global approaches to envenomation and treatments, handbook of venoms and toxins of reptiles* (2nd ed.). CRC Press.

Murthy, T. S. N. (1998). The venom system of Indian snakes. In B. D. Sharma (Ed.), *Snakes in India - A Source Book* (pp. 75–81). Asiatic Publishing house.

Nirthanan, S., Gopalakrishnakone, P., Gwee, M. C., Khoo, H. E., & Kini, R. M. (2003). Non-conventional toxins from Elapid venoms. *Toxicon, 41*, 397–407.

Ogawa, T., Nakashima, K., Oda, N., Shimohigashi, Y., Ohno, M., Hattori, S., Hattori, M., Sakaki, Y., & Kihara, H. (1995). *Trimeresurus flavoviridis* venom gland phospholipase A$_2$ isozymes genes have evolved via accelerated substitutions. *Journal of Molecular Recognition, 8*, 40–46.

Oh, A. M. F., Tan, C. H., Ariaranee, G. C., Quraishi, N., & Tan, N. H. (2017). Venomics of *Bungarus caeruleus* (Indian krait): Comparable venom profiles, variable immunoreactivities among specimens from Sri Lanka, India and Pakistan. *Journal of Proteomics, 164*, 1–18.

Ownby, C. (1990). Locally acting agents: myotoxin, haemorrhagic toxin and dermonecrotic factors. In W. T. Shier & D. Mebs (Eds.), *Handbook of toxinology* (pp. 602–654). Marcel Dekker.

Parveen, G., Khan, M. F., Ali, H., Ibrahim, T., & Shah, R. (2017). Determination of lethal dose (LD$_{50}$) of venom of four different poisonous snakes found in Pakistan. *Biochemistry and Molecular Biology, 3*, 18.

Pathan, J., Mondal, S., Sarkar, A., & Chakrabarty, D. (2017). Daboialectin, a C-type lectin from Russell's viper venom induces cytoskeletal damage and apoptosis in human lung cancer cells in vitro. *Toxicon, 127*, 11–21.

Patra, A., Kalita, B., Chanda, A., & Mukherjee, A. K. (2017). Proteomics and antivenomics of *Echis carinatus carinatus* venom: Correlation with pharmacological properties and pathophysiology of envenomation. *Nature Scientific Reports, 7*, 17119.

Patra, A., Chanda, A., & Mukherjee, A. K. (2019). Quantitative proteomic analysis of venom from Southern India common krait (*Bungarus caeruleus*) and identification of poorly immunogenic toxins by immune-profiling against commercial antivenom. *Expert Review of Proteomics, 16*(5), 457–469.

Patra, A., & Mukherjee, A. K. (2020). Proteomic analysis of Sri Lanka *Echis carinatus* venom: Immunological cross-reactivity and enzyme neutralization potency of Indian polyantivenom. *Journal of Proteome Research, 19*(8), 3022–3032.

Pierre, L. S., Flight, S., Masci, P. P., Hanchard, K. J., Lewis, R. J., Alewood, P. F., & Lavin, M. F. (2006). Cloning and characterisation of natriuretic peptides from the venom glands of Australian elapids. *Biochimie, 88*, 1923–1931.

Ponnappa, K. C., Saviour, P., Ramachandra, N. B., Kini, R. M., & Gowda, T. V. (2008). INN-toxin, a highly lethal peptide from the venom of Indian cobra (*Naja naja*) venom-Isolation, characterization and pharmacological actions. *Peptides, 29*(11), 1893–1900.

Prasad, B. N., Uma, B., Bhatt, S. K., & Gowda, V. T. (1999). Comparative characterisation of Russell's viper (*Daboia/Vipera russelii*) venoms from different regions of the Indian peninsula. *Biochimica et Biophysica Acta, 1428*(2–3), 121–136.

Pukrittayakamee, S., Warrell, D. A., Desakorn, V., McMichael, A. J., White, N. J., & Bunnag, D. (1988). The hyaluronidase activities of some southeast Asian snake venoms. *Toxicon, 26*, 629–637.

Pung, Y. F., Wong, P. T. H., Kumar, P. P., Hodgson, W. C., & Kini, R. M. (2005). Ohanin, a novel protein from king cobra venom induces hypolocomotion and hyperalgesia in mice. *The Journal of Biological Chemistry, 280*, 13137–13147.

Raba, R., Aaviksaar, A., Raba, M., & Siigur, J. (1979). Cobra venom acetylcholinesterase. *European Journal of Biochemistry, 96*, 151–1580.

Rajagopalan, N., Pung, Y. F., Zhu, Y. Z., Wong, P. T., Kumar, P. P., & Kini, R. M. (2007). Beta-cardiotoxin: a new three-finger toxin from *Ophiophagus hannah* (king cobra) venom with beta-blocker activity. *FASEB Journal, 21*(13), 3685–3695.

Reyes-Velasco, J., Card, D. C., Andrew, A. L., Shaney, K. J., Adams, R. H., Schield, D. R., Casewell, N. R., Mackessy, S. P., & Castoe, T. A. (2015). Expression of venom gene homologs in diverse python tissues suggests a new model for the evolution of snake venom. *Molecular Biology and Evolution, 32*(1), 173–183.

Ribeiro, P. H., Zuliani, J. P., Fernandes, C. F., Calderon, L. A., Stábeli, R. G., Nomizo, A., & Soares, A. M. (2016). Mechanism of the cytotoxic effect of l-amino acid oxidase isolated from *Bothrops alternatus* snake venom. *International Journal of Biological Macromolecules, 92*, 329–337.

Richards, D. P., Barlow, A., & Wüster, A. (2011). Venom lethality and diet: Differential responses of natural prey and model organisms to the venom of the saw-scaled vipers (*Echis*). *Toxicon, 59* (1), 110–116.

Richards, D. P., Barlow, A., & Wüster, W. (2012). Venom lethality and diet: Differential responses of natural prey and model organisms to the venom of the saw-scaled vipers (*Echis*). *Toxicon, 59*, 110–116.

Risch, M., Georgieva, D., Bergen, M., Jehmlich, N., Genov, N., Arni, R. K., & Betzel, C. (2009). Snake venomics of the Siamese Russell's viper (*Daboia russelii siamensis*)—Relation to pharmacological activities. *Journal of Proteomics, 72*, 256–269.

Rosenberg, H. (1967). Histology, histochemistry, and emptying mechanism of the venom glands of some elapid snakes. *Journal of Morphology, 123*, 133–156.

Sachidananda, M. K., Murari, S. K., & Channe Gowda, D. (2007). Characterization of an antibacterial peptide from Indian cobra (*Naja naja*) venom. *Journal of Venomous Animals and Toxins Including Tropical Diseases, 13*, 446–461.

Saikia, D., Thakur, R., & Mukherjee, A. K. (2011). An acidic phospholipase A_2 (RVVA-PLA$_2$-I) purified from *Daboia russelii* venom exerts its anticoagulant activity by enzymatic hydrolysis of plasma phospholipids and by non-enzymatic inhibition of factor Xa in a phospholipids/Ca^{2+} independent manner. *Toxicon, 57*, 841–850.

Saikia, D., Bordoloi, N. K., Chattopadhyay, P., Chocklingam, S., Ghosh, S. S., & Mukherjee, A. K. (2012). Differential mode of attack on membrane phospholipids by an acidic phospholipase A_2 (RVVA-PLA$_2$-I) from *Daboia russelii* venom. *Bichim Biophys Acta –Biomembrane, 12*, 3149–3157.

Saikia, D., Majumdar, S., & Mukherjee, A. K. (2013). Mechanism of *in vivo* anticoagulant and haemolytic activity by a neutral phospholipase A_2 purified from *Daboia russelii russelii* venom: Correlation with clinical manifestations in Russell's viper envenomed patients. *Toxicon, 76*, 291–300.

Saikia, S., & Mukherjee, A. K. (2017). Anticoagulant and membrane damaging properties of snake venom phospholipase A_2 enzymes. In P. Gopalakrishnakone, H. Inagaki, A. K. Mukherjee, T. R. Rahmy, & C. W. Vogel (Eds.), *Handbook of toxinology, volume – snake venom* (pp. 87–104). Springer Nature.

Sales, P. B. V., & Santoro, M. L. (2008). Nucleotidase and DNase activities in Brazilian snake venoms. *Comparative Biochemistry and Physiology, 147*, 85–95.

Saviola, A. J., Chiszar, D., Busch, C., & Mackessy, S. P. (2013). Molecular basis for prey relocation in viperid snakes. *BMC Biology, 11*(1), 20.

Saviola, A. J., Modahl, C. M., & Mackessy, S. P. (2015). Disintegrins of *Crotalus simus tzabcan* venom: Isolation, characterization and evaluation of the cytotoxic and anti-adhesion activities of tzabcanin, a new RGD disintegrin. *Biochimie, 116*, 92–102.

Saviola, A. J., Burns, P. D., Mukherjee, A. K., & Mackessy, S. P. (2016). The disintegrin Tzabcanin inhibits adhesion and migration in melanoma and lung cancer cells. *International Journal of Biological Macromolecules, 88*, 457–464.

Senji Laxme, R. R., Khochare, S., de Souza, H. F., Ahuja, B., Suranse, V., Martin, G., Whitaker, R., & Sunagar, K. (2019). Beyond the 'big four': Venom profiling of the medically important yet neglected Indian snakes reveals disturbing antivenom deficiencies. *PLoS Neglected Tropical Diseases, 13*, e0007899.

Shashidharamurthy, R., Mahadeswaraswamy, Y. H., Ragupathi, L., Vishwanath, B. S., & Kemparaju, K. (2010). Systemic pathological effects induced by cobra (*Naja naja*) venom from geographically distinct origins of Indian peninsula. *Experimental and Toxicologic Pathology, 62*(6), 587–592.

Sharma, M., Das, D., Iyer, J. K., Kini, R. M., & Doley, R. (2015). Unveiling the complexities of *Daboia russelii* venom, a medically important snake of India, by tandem mass spectrometry. *Toxicon, 107*, 266–281.

Sharma, M., Iyer, J. K., Shih, N., Majumder, M., Mattaparthi, V. S., Mukhopadhyay, R., & Doley, R. (2016). Daboxin P, a major Phospholipase A_2 enzyme from the Indian *Daboia russelii russelii* venom targets Factor X and Factor Xa for its anticoagulant activity. *PLoS One, 11*, e0153770.

Shimokawa, K.-I., & Takahashi, H. (1995). Comparative study of fibrinogen degradation by four arginine ester hydrolases from the venom of *Agkistrodon caliginosus* (Kankoku-Mamushi). *Toxicon, 33*, 179–186.

Sintiprungrat, K., Watcharatanyatip, K., Senevirathne, W. D., Chaisuriya, P., Chokchaichamnankit, D., Srisomsap, C., & Ratanabanangkoon, K. A. (2016). Comparative study of venomics of *Naja naja* from India and Sri Lanka, clinical manifestations and antivenomics of an Indian polyspecific antivenom. *Journal of Proteomics, 132*, 131–143.

Sridharan, S., Kini, R. M., & Richards, A. M. (2020). Venom natriuretic peptides guide the design of heart failure therapeutics. *Pharmacological Research, 155*, 104687.

Stocker, K. F. (1990a). Composition of snake venoms. In K. F. Stocker (Ed.), *Medical use of snake venom proteins* (pp. 33–56). CRC Press.

Taborsk, E., & Kornalik, F. (1985). Individual variability of *Bothrops asper* venom. *Toxicon, 23*, 612.

Teng, C. M., Chen, Y. H., & Ouyang, C. (1984). Purification and properties of the main coagulant and anticoagulant principles of *Vipera russelii* snake venom. *Biochimica et Biophysica Acta, 786*, 204–212.

Thakur, R., Kumar, A., Bose, B., Panda, D., Saikia, D., Chattopadhyay, P., & Mukherjee, A. K. (2014). A new peptide (Ruviprase) purified from the venom of *Daboia russelii russelii* shows potent anticoagulant activity via non-enzymatic inhibition of thrombin and factor Xa. *Biochimie, 105*, 149–158.

Thakur, R., Chattopadhyay, D., Ghosh, S. S., & Mukherjee, A. K. (2015). Elucidation of procoagulant mechanism and pathophysiological significance of a new prothrombin activating metalloprotease purified from *Daboia russelii russelii* venom. *Toxicon, 100*, 1–12.

Thakur, R., Kini, S., Karkulang, S., Banerjee, A., Chatterjee, P., Chanda, A., Chatterjee, A., Panda, D., & Mukherjee, A. K. (2016). Mechanism of apoptosis induction in human breast cancer MCF-7 cell by Ruviprase, a small peptide from *Daboia russelii russelii* venom. *Chemico-Biological Interactions, 258*, 297–304.

Thakur, R., & Mukherjee, A. K. (2017a). A brief appraisal on Russell's viper venom (*Daboia russelii*) proteinases. In P. Gopalakrishnakone, H. Inagaki, A. K. Mukherjee, T. R. Rahmy, & C. W. Vogel (Eds.), *Handbook of toxinology, Volume– snake venom* (pp. 123–144). Springer Nature. https://doi.org/10.1007/978-94-007-6648-8_18-1

Thakur, R., & Mukherjee, A. K. (2017b). Pathophysiological significance and therapeutic applications of snake venom protease inhibitors. *Toxicon, 131*, 37–47.

Thakur, R., Chattopadhyay, P., & Mukherjee, A. K. (2019). The wound healing potential of a pro-angiogenic peptide purified from Indian Russell's Viper (*Daboia russelii*) venom. *Toxicon, 165*, 72–82.

Trummal, K., Tõnismägi, K., Paalme, V., Järvekülg, L., Siigur, J., & Siigur, E. (2011). Molecular diversity of snake venom nerve growth factors. *Toxicon, 58*, 363–368.

Weinstein, S. A., Smith, T. L., & Kardong, K. V. (2009). Reptile venom glands form, function, and future. In *Handbook of venoms and toxins of reptiles* (pp. 65–91). CRC Press.

Saha, A., Gomes, A., Chakravarty, A. K., Biswas, A. K., Giri, B., Dasgupta, S. C., & Gomes, A. (2006). CNS and anticonvulsant activity of a non-protein toxin (KC-MMTx) isolated from King Cobra (*Ophiophagus hannah*) venom. *Toxicon, 47*, 296–303.

Shibata, H., Chijiwa, T., Oda-Ueda, N., et al. (2019). The habu genome reveals accelerated evolution of venom protein genes. *Scientific Reports, 8*, 11300.

Sridharan, S., & Kini, R. M. (2015). Snake venom natriuretic peptides: Potential molecular probes. *BMC Pharmacology and Toxicology, 16*(Suppl 1), A87.

Stocker, K. F. (1990b). Composition of snake venoms. In K. F. Stocker (Ed.), *Medical use of snake venom proteins* (pp. 33–56). CRC Press.

Sunagar, K., Jackson, T. N. W., Reeks, T., & Fry, B. G. (2015). *Cysteine-rich secretory proteins, venomous reptiles and their toxins: Evolution, pathophysiology and biodiscovery* (pp. 239–246). Oxford University Press.

Sundell, I. B., Rånby, M., Zuzel, M., Robinson, K. A., & Theakston, R. D. G. (2003). In vitro procoagulant and anticoagulant properties of *Naja naja naja* venom. *Toxicon, 42*, 239–247.

Tan, K. K., Bay, B. H., & Gopalakrishnakone, P. (2018). L-amino acid oxidase from snake venom and its anticancer potential. *Toxicon, 144*, 7–13.

Thomas, R. B., & Pough, F. H. (1979). The effect of rattlesnake venom on digestion of prey. *Toxicon, 17*, 221–228.

Tsai, I. H., Lu, P. J., & Su, J. C. (1996). Two types of Russell's viper revealed by variation in phospholipases A_2 from venom of the subspecies. *Toxicon, 34*(1), 99–109.

Valente, R. H., Luna, M. S., de Oliveira, U. C., Nishiyama-Junior, M. Y., Junqueira-de-Azevedo, I. L., Portes-Junior, J. A., Clissa, P. B., Viana, L. G., Sanches, L., Moura-da-Silva, A. M., Perales, J., & Yamanouye, N. (2018). *Bothrops jararaca* accessory venom gland is an ancillary source of toxins to the snake. *Journal of Proteomics, 177*, 137–147.

Vivas-Ruiz, D. E., Gonzalez-Kozlova, E. E., Delgadillo, J., Sandoval, G. A., Lazo, F., Rodríguez, E., Yarlequé, A., & Sanchez, E. F. (2019). Biochemical and molecular characterization of the hyaluronidase from *Bothrops atrox* Peruvian snake venom. *Biochimie, 162*, 33–45.

Vidal, N. (2002). Colubroid systematics: evidence for an early appearance of the venom apparatus followed by extensive evolutionary tinkering. *Journal of Toxicology – Toxin Reviews, 21*(1–2), 21–41.

Vogel, C. W., & Fritzinger, D. C. (2010). Cobra venom factor: structure, function, and humanization for therapeutic complement depletion. *Toxicon, 56*, 1198–1222.

Warrell, D. A. (1989). Snake venoms in science and clinical medicine. 1. Russell's viper: biology, venom and treatment of bites. *Transactions of the Royal Society of Tropical Medicine and Hygiene, 83*(6), 732–740.

Yamazaki, Y., & Morita, T. (2004). Structure and function of snake venom cysteine-rich secretory proteins. *Toxicon, 44*, 227–231.

Zeller, E. A. (1948). Enzymes of snake venoms and their biological significance. In F. F. Nord (Ed.), *Advances in enzymology* (pp. 459–495). Inderscience Publishers.

Zeller, E. A. (1950). The formation of pyrophosphate from adenosine triphosphate in the presence of a snake venom. *Archives of Biochemistry, 28*, 138–139.

Zhang, Y., Xu, X., Shen, D., Song, J., Guo, M., & Yan, X. (2012). Anticoagulation factor I, a snaclec (snake C-type lectin) from Agkistrodon acutus venom binds to FIX as well as FX: Ca2+ induced binding data. *Toxicon, 59*, 718–723.

Indian Spectacled Cobra (*Naja naja*)

4

Abstract

The Indian cobra (*Naja naja*), which is popularly known as the Asian cobra, is also called the spectacled cobra because of its hood mark. It inhabits the Indian subcontinent but is also found in the northeastern parts of the country. The Indian cobra has received much respect and attention in Indian mythology. The bite of this species of snake requires immediate medical attention. Another species of Indian cobra (known as the Indian monocled cobra or *N. kaouthia*), found in eastern and northeastern parts of the country, is also a deadly venomous snake, though antivenom against this species is not produced commercially. Indian cobras can grow 1–1.5 m in length and show great variation in color. The Indian cobra has proteroglyph dentition: the two hollow, short front fangs are fixed to the top jaw at the front of the mouth. The Indian cobra is oviparous, laying between 10 and 40 eggs. The biochemical analysis of the composition of Indian cobra venom shows that it is mainly comprised of nonenzymatic toxins, especially three-finger toxins, and it also contains enzymatic toxins of variable composition depending on the zoogeographic origin of the cobras. Recently, the relative proportion of different toxins in Indian cobra venom from different regions of the country has been deciphered by proteomic analysis and the variation in clinical manifestation post-cobra bite has also been demonstrated. The species-specific differences in venom composition between *N. naja* and *N. kaouthia*, based on biochemical and proteomic analyses, are also described in detail in this chapter. Because of the variations in venom composition, venom samples from these two closely related Indian cobra species also have different pharmacological properties and toxicities.

Keywords

Andaman cobra · Central Asian cobra · Clinical features of cobra envenomation · Cobra venom toxins · Differences in venom composition · Indian cobra venom composition · Indian monocled cobra · Indian spectacled cobra · Intrazonal variation · *Naja kaouthia* · *Naja naja* · *Naja oxiana* · Neurotoxicity · Omics analyses of cobra venom toxins

4.1   Taxonomic Classification of the Indian Spectacled Cobra (*Naja naja*)

Phylum: Chordata
Group: Vertebrata
Subphylum: Gnathostomata
Class: Reptilia
Subclass: Diapsida
Order: Squamata
Suborder: Ophidia
Infraorder: Caenophidia
Family: Elapidae
Genus: *Naja*
Species: *naja*

4.2   Characteristic Features of the Indian Spectacled Cobra

The Indian cobra (*N. naja*), popularly known as the Asian cobra, is often called the spectacle cobra because of its hood mark (Fig. 1.3a and Fig. 4.1). It belongs to the Elapidae family of snakes and is one of the "Big Four" venomous snakes of India, with a bite that demands immediate medical attention.

Indian cobras have received much respect and attention in Indian culture. In Hindu mythology, the spectacle is believed to be the footmark of Lord Vishnu. The snake is the garland of Lord Shiva. Several popular Indian films, particularly the Bollywood Hindi movies, are based on stories related to this snake, with its distinctive and impressive hood feature. When the cobra is threatened, it raises the front part of its body and stretches some ribs in its neck region, which takes the shape of a hood. This snake is an endangered species in India and it is protected under the Indian Wildlife Protection Act (1972). Local names for the Indian cobra are shown in Table 4.1.

Josephus Nicolaus Laurenti in 1768 described the genus of this snake; however, Swedish zoologist and botanist Carl Linnaeus in 1758 was the first to describe *N. naja* (Linnaeus, 1758). Other species of *Naja* (shown below) are also distributed across the country and contribute to snakebite mortality in different locales of India (Wüster, 1998). Nevertheless, these species are not covered under the "Big Four" venomous snakes of India and commercial polyvalent antivenom does not contain antibodies against the other species of cobra.

Fig. 4.1 Photographs of the Indian spectacled cobra (*Naja naja*) in an aggressive disposition (PC Vivek Sharma)

1. The Indian monocled cobra or Bengal cobra (*N. kaouthia*) (Fig. 4.2) is found in West Bengal, Odisha, Northeast India, and Bangladesh. Biochemical and proteomic analyses have uncovered the venom composition of *N. naja* (Chanda, Kalita, et al., 2018a) and revealed differences in the venom composition between *N. naja* and *N. kaouthia* (Mukherjee & Maity, 2002; Chanda, Patra, et al., 2018b).
2. The central Asian cobra or black cobra (*N. oxiana*) is distributed in northwest India and Pakistan. This species can be differentiated from all other Asiatic cobras on the basis of its high ventral and/or subcaudal scale counts (Wüster & Thorpe, 1989). This species has a dark but never fully black color, and unlike *N. naja* and *N. kaouthia* it does not have the hood mark (Fig. 4.3a).
3. The Andaman cobra (*N. sagittifera*) is endemic to the Andaman Islands of India (Wüster, 1998). Little information is available about this species of cobra (Fig. 4.3b,c).

The average length of the Indian cobra (*N. naja*) is 1.0 m (3.3 ft), but it can grow to a maximum length of 1.5 m (4.9 ft). The Indian cobra possesses smooth scales with black eyes, a broad neck, and a medium body size (Whitaker, 2006). A

Table 4.1 Vernacular names for the Indian spectacled cobra

Language	Local name
Assamese	Phetigom
Bengali	Gokhra (গোখরো), khoris (খরিস)
Hindi	Nag (नाग)
Gujarati	Nag (નાગ)
Kannada	Nagara Haavu (ನಾಗರ ಹಾವು)
Malayalam	Moorkhan (മൂർഖൻ)
Marathi	Nag (नाग)
Odia	(ଗୋଖର ସାପ/ନାଗ ସାପ)
Tamil	Nalla pambu (நல்ல பாம்பு)
Telegu	Nagu Paamu (నాగు పాము)

Fig. 4.2 Photographs of the Indian monocled cobra (*Naja kaouthia*) (PC Vivek Sharma)

fascinating variation in color can be seen in the Indian cobra (*N. naja*) across the country, from black to dark brown, tan, reddish, and yellowish (Whitaker et al., 2004; Whitaker, 2006). The Indian cobra is characterized by its display of a hood mark on its dorsal scales, and posteriorly, convex light bands occur at the 20th to 25th ventral scales where two circular ocelli patterns are connected by a curved line. The pattern arouses an image of spectacles in the rear head (dorsal side) of the cobra, which is its characteristic feature. In Indian mythology, the hood mark is believed to be a footprint of Lord Krishna. When threatened or provoked, the snake raises its forebody, expands its large impressive hood, and produces a loud hissing sound that is easily identified (Fig. 4.1). If an enemy or human does not take the warning seriously, the snake has no option other than biting. The head of *N. naja* is elliptical in shape, depressed, and marginally distinct from its neck. The eyes are medium with round pupils. The head of the cobra possesses distinct scales (Vijayaraghavan, 2008). The supralabial scale, the third scale on the upper lip, touches the eye and nasal area. Another pair of small wedge-shaped scales (the infralabial scales) are found between the fourth and fifth scales on either side of the lower lip (Vijayaraghavan, 2008).

The Indian cobra has proteroglyph dentition: the two hollow, short front fangs of the cobra are fixed to the top jaw at the front of the mouth (Bogert, 1943; Wüster &

a b c

Fig. 4.3 (**a**) Photograph of *N. oxiana* (source of photo: Omid Mozaffari/CC BY-SA) (https://creativecommons.org/licenses/by-sa/3.0). (**b, c**) Photographs of the Andaman cobra (*N. sagittifera*) (source of photo: reptile-database.reptarium, and http://www.indianreptiles.org/sp/436/Naja-sagittifera, PC S. Harikrishnan)

Thorpe, 1992). The fangs of *N. naja*, in comparison to Russell's viper fangs, are shorter (Wüster & Thorpe, 1992). The venom is injected into the prey or victim via the hollow fangs and the amount of venom, which depends on the size of the victim, is believed to be under muscular control (Kochva, 1987; Weinstein et al., 2009). Cobras have a tremendous sense of smell and night vision that help them to capture prey. Lateral views of the elapid skull are shown in Fig. 4.4.

4.3 Geographical Distribution and Reproduction of the Indian Spectacled Cobra

The Indian cobra inhabits the Indian subcontinent, including areas in Nepal, Bangladesh, India, Pakistan, and Sri Lanka. This species is less abundant in Northeast India where *N. kaouthia* predominates. It is found in the plains, wetlands, dense or open jungles, and open fields, near bodies of water and in heavily populated urban regions (Fig. 4.5). Its distribution is spread from sea level to an elevation of 6600 ft. (2000 m). These snakes are commonly found near rice farms where an abundance of rats can be found, which serve as food for the snakes. They prefer staying in mammalian holes, termite mounds, and tree hollows.

The Indian cobra is oviparous, laying 10–40 eggs between April and July. The mother hatches and protects the eggs and young babies (hatchlings), which are 8–12 in. in length and have fully functioning venom glands and fangs. The fangs come out from the eggs at about 3 months post-laying of eggs, so that even the young snakes can inject lethal venom into victims.

Fig. 4.4 Left lateral view of an elapid skull. Several short, solid maxillary teeth occur after the fangs. Abbreviations: *bc* braincase, *ec* ectopterygoid, *f* fang, *man* mandible, *mx* maxilla, *n* nasal, *pal* palatine, *pf* prefrontal, *pmx* premaxilla, *pt.* pterygoid, *q* quadrate, *smx* septomaxilla, *st* supratemporal (reprinted with permission from Deufel & Cundall, 2006)

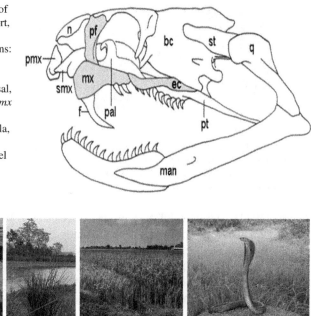

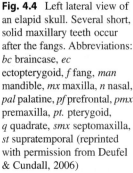

Fig. 4.5 Natural habitats of the Indian cobra. (**a**) Tree hollows, (**b**) brickfield, (**c**) wetlands, (**d**) near rice plantations, (**e**) open grassland (photos **a–d** courtesy of the author and e courtesy of Vivek Sharma)

4.4   Biochemical Composition of the Indian Spectacled Cobra Venom

Indian cobra (*N. naja*) venom is highly soluble in water and aqueous buffers. The aqueous solution of *N. naja* venom is pale yellowish in color, with a relative viscosity of 1.03–1.07, and a somewhat acidic pH (pH range from 6.6 to 7.0) (Devi, 1968). The Indian cobra (*N. naja*) venom is composed of approximately 90–95% protein and polypeptides (Mukherjee, 1998); the toxins either individually or in the form of cognate complexes of toxins show the pharmacological effects of envenomation (Meier & Stocker, 1995; Dutta et al., 2019). Apart from the proteins and polypeptides, *N. naja* venom contains approximately 5% non-proteinaceous, usually low-molecular-weight organic and inorganic substances, such as amino acids, amines, lipids, nucleosides and nucleotides, carbohydrates, and metal ions (Table 4.2). Most of the non-proteinaceous substances do not show pharmacological activity but they may augment the toxicity of venom proteins or may be required for marinating the stability and function of several venom toxins (Bieber, 1979). Further studies are required to decipher the exact biological role(s) of the nonprotein substances of snake venom.

Table 4.2 Inorganic and organic constituents of an Indian spectacled cobra (*N. naja*) venom sample obtained from the Burdwan district of West Bengal, eastern India

Parameters	Values
Total nitrogen (%)	20 ± 1.8
Total phosphorus (µg/100 mg venom)	18.6 ± 0.9
Inorganic phosphorus (µg/100 mg venom)	10.8 ± 1.1
Acid-soluble phosphorus (µg/100 mg venom)	7.9 ± 1.1
Chloride (µg/100 mg venom)	60 ± 3.2
Li^+ (µg/100 mg venom)	102 ± 3.3
Na^+ (mg/100 mg venom)	3.6 ± 0.4
K^+ (µg/100 mg venom)	76.5 ± 2.1
Ca^{2+} (µg/100 mg venom)	61 ± 1.9
Total lipid (mg/100 mg venom)	5.6 ± 0.8
Total carbohydrate (µg/100 mg venom)	1.9 ± 0.05
Total protein (%)	96 ± 1.1
(a) Albumin (% of total protein)	70.3 ± 1.8
(b) Globulin ((% of total protein)	29.7 ± 1.9
Free amino acids	Nil
Ribose content (mg/100 mg venom)	0.8 ± 0.03

Values are mean ± SD of triplicate determinations (cited in Mukherjee, 1998)

The proteins and polypeptides in snake venom belong to approximately 26 different snake venom protein families, depending on the species and geographical origin. The venom of *N. naja* from different regions does not necessarily contain all of the enzymes and toxins. Irrespective of the variations, 12 venom protein families have been reported to be common to all snake venoms, each making a distinct functional contribution to the pathophysiology of envenomation (Bieber, 1979; Mebs, 1985). Venom proteins, including *N. naja* venom toxins, are broadly classified into two groups: (a) the enzymatic proteins and (b) the nonenzymatic toxins (proteins/polypeptides) (Koh et al., 2006; McCleary & Kini, 2013). Proteomic analysis has demonstrated that, depending on the geographical origin of *N. naja*, its venom contains 38–61% low-molecular-mass (generally <10 kDa) nonenzymatic three-finger toxins (3FTxs) that have a wide array of pharmacological actions in *N. naja* envenomation (Dutta et al., 2017; Chanda et al., 2018a,b).

Because of the predominating 3FTxs, which exhibit innumerable clinical symptoms such as neurotoxicity, cardiotoxicity, cytotoxicity, and antiplatelet effects (Fry et al., 2003; Kini & Doley, 2010), the Indian cobra (*N. naja*) venom shows neurotoxicity and neuroparalytic symptoms in envenomed patients and experimental animals (Mukherjee, 2020), as discussed below. Besides the 3FTxs, the other nonenzymatic toxins of Indian cobra venom are cystatin, cysteine-rich secretory proteins (CRISP), cobra venom factors (CVF), Kunitz-type serine protease inhibitors (KSPI), natriuretic peptides (NP), nerve growth factors (NGF), Ohanin-like proteins (OLP) or vespryns, snaclecs or snake venom C-type lectins, and vascular endothelial growth factors (VEGF). Nevertheless, less than 5% relative abundance is reported for the individual minor components of Indian cobra (*N. naja*) and Indian monocled cobra (*N. kaouthia*) venom (Dutta et al., 2017, Chanda et al., 2018a, b; reviewed by Kalita & Mukherjee, 2019; Chanda & Mukherjee, 2020a).

Phospholipase A_2 (PLA_2), which catalyzes the phospholipid hydrolysis and also shows an anticoagulant activity, is the most abundant enzymatic protein in Indian cobra (*N. naja*) venom (Dutta et al., 2015, 2017; Chanda et al., 2018a, b). The other enzymatic proteins of this venom are aminopeptidases, cholinesterase (including acetylcholinesterase and butyryl cholinesterase), hyaluronidases, L-amino acid oxidases (LAAO), $5'$-nucleotidases, phosphodiesterases, phospholipase Bs (PLB), snake venom metalloprotease, and serine proteases (SVMP/SVSP) (Dutta et al., 2017; Choudhury et al., 2017; Chanda et al., 2018a, b; Chanda & Mukherjee, 2020b; Mukherjee, 2020).

Various enzymes and nonenzymatic proteins (toxins) have been purified and characterized from Indian cobra (*N. naja*) venom, and their biological activities are shown in Tables 4.3 and 4.4.

Separation of Indian cobra (*N. naja*) venom proteins (toxins) by SDS-PAGE under reduced and non-reduced conditions demonstrates their differential migrations, which suggests the occurrence of multiple subunits, noncovalent oligomers (multimeric forms), self-aggregation of proteins, and/or interactions among cobra venom proteins (Chanda et al., 2018a, 2018b) to form cognate protein complexes in a definite stoichiometric ratio. This supplements the pharmacological activity and toxicity of individual toxins of cobra venom (Dutta et al., 2019). Two types of toxin synergisms have been proposed to augment the toxicity and/or pharmacological activity of individual toxins of a complex: (a) intermolecular synergism and (b) supramolecular synergism (Laustsen, 2016). Intermolecular synergism refers to a condition when two or more interacting toxins have two or more targets on one or more related biological pathways that synergistically increases their toxicity (Laustsen, 2016). Supramolecular synergism denotes a condition where the venom toxins form stable or cognate complexes that synergistically enhance the toxicity of individual toxins of the complex so that the interacting toxins create a hyper-potentiated toxin (Laustsen, 2016). The best example of supramolecular synergism is the complex formation of cobra venom cytotoxins, where cytotoxin enhances PLA_2 activity that increases the hydrolysis of cell membranes to destabilize the system (Mukherjee, 2008; Mukherjee, 2010; Chaim-Matyas et al., 1995; Dutta et al., 2019). The acidic phospholipase A_2 cognate complex has a molecular mass of 20–24 kDa and it consists of an acidic phospholipase A_2 enzyme (NnPLA$_2$-I), 3FTxs (cytotoxin and neurotoxin), and a trace quantity of nerve growth factor, which causes cell cytotoxicity after binding to vimentin of rat myoblast L6 cells. Individual toxins are unable to show cytotoxicity (Dutta et al., 2019).

Table 4.3 Enzymatic proteins purified and characterized from Indian cobra (*N. naja*) venom

Enzyme class	Name of enzyme	Molecular weight (kDa)	Biological functions	References
PLA_2	NN-XIa-PLA_2	10.7	Myotoxicity, edema, and neurotoxicity in mice, LD_{50} (i.p.) of 8.5 mg/kg body weight of mice. Also shows antimicrobial activity against the human pathogenic strains	Basavarajappa and Gowda (1992), Machiah and Gowda (2006), Sudarsan and Dhananjaya (2015)
	NN-XIb-PLA2s	15.0	It induces acute neurotoxicity in mice, LD_{50} (i.p.) of 0.22 mg/kg body weight of mice	
	NN-XIII-PLA_2	11.2	Higher toxicity to mice compared to crude venom, LD_{50} (i.p.) of 2.4 mg/kg body weight of mice	Bhat and Gowda (1989)
	NN-IVb1-PLA_2	11–11.5	Shows neurotoxicity in mice but does not show myotoxicity, anticoagulant, edema-inducing, and direct hemolytic activities	Bhat and Gowda (1991)
	NN-Vb-PLA_2	10.5–11.0	Displays neurotoxic symptoms in mice but does not show myotoxicity, anticoagulant, and edema-inducing activities. LD_{50} is 0.27 mg/kg	Bhat et al. (1991)

(continued)

Table 4.3 (continued)

Enzyme class	Name of enzyme	Molecular weight (kDa)	Biological functions	References
			body weight of mice	
	Acidic PLA$_2$ (PL-1a, PL-1b, and PL-1c)	Not mentioned	Not mentioned	Jayanthi and Gowda (1983)
	Acidic phospholipases A$_2$ isoenzymes (NN-I2c-PLA$_2$, NN-I2d-PLA$_2$, and NN-I2c-PLA$_2$)	13–14.5	These isoenzymes induce edema in mouse footpads and demonstrate cytotoxicity against Ehrlich ascites tumor cells	Rudrammaji and Gowda (1998)
	NND-IV-PLA ($_2$)	13.262	Inhibits the ADP, collagen, and epinephrine-induced platelet aggregation, class B1 platelet aggregation inhibitor	Satish et al. (2004)
	Nn-PLA$_2$-I	15.2 (SDS-PAGE), 14.186 (MALDI-ToF-MS)	Anticoagulant activity against platelet-poor plasma, thrombin inhibitor, causes dose-dependent deaggregation of platelet-rich plasma (PRP), inhibits the collagen- and thrombin-induced aggregation of PRP. However, this enzyme does not show in vitro cell cytotoxicity, bactericidal activity, and	Dutta et al. (2015)

(continued)

Table 4.3 (continued)

Enzyme class	Name of enzyme	Molecular weight (kDa)	Biological functions	References
			hemolytic activity	
SVMP	NN-PF3	67.81	Inhibition of collagen-induced aggregation of human platelets by binding to integrin α2β1; myotoxicity, cytotoxicity, and hemorrhagic activity. Does not show myotoxicity, cytotoxicity, and hemorrhagic activity but anticoagulant in nature, nonlethal to mice at a dose of 15 mg/kg body weight	Jagadeesha et al. (2002), Kumar et al. (2011)
LAAO	LAAO	65.0	Not described	Neema et al. (2015)
Hyaluronidase	NNH1	70.406	In an indirect manner potentiates the myotoxicity of VRV-PL-VIII, a myotoxin, and the hemorrhagic potency of a hemorrhagic complex I of cobra venom, in vivo degradation of hyaluronan in the extracellular matrix (ECM) and therefore it acts as "spreading factor"	Girish, Mohanakumari, et al. (2004a), Girish, Shashidharamurthy, et al. (2004b)

Table 4.4 Nonenzymatic proteins purified and characterized from Indian cobra (*N. naja*) venom

Nonenzymatic protein class	Name of nonenzymatic proteins	Molecular weight (kDa)	Biological functions	References
KSPI	Chymotrypsin Kunitz inhibitor type of polypeptide	6.2	Strongly inhibits trypsin, other biological functions are unknown	Shafqat et al. (1990)
Cytotoxins or cytotoxin-like proteins	CLBP	~6.7	Not described	Babu et al. (1995)
	LCBP		The authors stated that "the cytotoxicity toward Yoshida sarcoma cells and lethal toxicity toward mice of LCBP were both one order of magnitude lower than that of cytotoxins and that of toxin A, respectively"	Takechi et al. (1987)
	NN-32	6.7	Demonstrates in vitro cytotoxicity on EAC cells, shows cardiotoxicity on isolated guinea pig auricle, displays antioxidant property; anti-arthritic and anti-inflammatory activity in mice	Das et al. (2011)
	Cytotoxin isoforms (CTX2, CTX7, CTX8, CTX9, CTX10)	6.71–8.04	In vitro hemolytic and cytotoxic activities by perturbing the membrane phospholipids	Suzuki-Matsubara et al. (2016)
	Cytotoxin 2a	6.74	Demonstrated in vitro cytotoxicity toward Yoshida sarcoma and ascites hepatoma cells	Kaneda et al. (1984)
Neurotoxin	Cobra venom neurotoxin	6.3	Neurotoxicity, LD_{50} (i.p.) of 0.2 mg/kg body weight of mice	Charles et al. (1981)
	Toxins B, C, D or long neurotoxins	~7.8	LD_{50} value of these toxins in experimental mice is about half compared to the neurotoxins from the elapid and hydrophid family of snake venoms; no other	Ohta et al. (1976, 1981a, 1981b)

(continued)

Table 4.4 (continued)

Nonenzymatic protein class	Name of nonenzymatic proteins	Molecular weight (kDa)	Biological functions	References
			biological activity is mentioned	
	Miscellaneous-type neurotoxin isoforms (weak neurotoxin)	~7.58	They contain 62–65 residues; however, biological properties are uncharacterized	Shafqat et al. (1991)
Cardiotoxin	Cardiotoxin isoforms (CTX1, CTX2, and CTX3)	6.74–6.79	Show in vitro hemolytic activity	Gorai and Sivaraman (2017)
Disintegrin	Disintegrin protein	64.0	Displays cytotoxicity against various types of human cancer cells	Thangam et al. (2012)
Cysteine-rich venom protein	CRVP1_NAJNA	3.9	No biological activity was determined	Suzuki et al. (2010)
Nerve growth factor	Cobra venom NGF	13.02	No biological activity was determined	Suzuki et al. (2010)
Unclassified proteins/ peptides	INN toxin	6.95	Exhibited neurotoxicity and cytotoxicity; however, does not inhibit cholinesterase activity, LD50 value 1.2 mg/kg body weight of mice	Ponnappa et al. (2008)
	Antihemorrhagic protein (NNAh)	~44.0	Hemorrhage and myonecrosis inhibitory activity	Suvilesh et al. (2017)
	Toxin A	6.3	LD_{50} is 0.15 mg/kg in mice. The tertiary structure resembles with cobra venom neurotoxins; however, no biological activity is demonstrated	Nakai et al. (1971)
	Antibacterial peptide	2.49	Potent antibacterial activity against Gram-positive and Gram-negative bacteria, no direct or indirect hemolytic activity	Sachidananda et al. (2007)

4.5   Biochemical and Proteomic Analyses to Demonstrate the Geographical Differences in Venom Composition of Indian Spectacled Cobra (*N. naja*) Venom

Before 2015, the venom composition of snakes of India was determined mainly by traditional biochemical analyses, which generally involved the initial separation of the venom proteins by a combination of liquid chromatographic techniques (i.e., gel filtration, ion exchange, and reversed-phase HPLC). The individual fractions were then assayed for enzyme activity and biological/pharmacological properties, and the proteins in each fraction and/or the crude venom were analyzed by sodium dodecyl sulfate (SDS)-polyacrylamide gel electrophoresis (SDS-PAGE). The chromatographic profiles of venom samples from the same species of Indian cobra but from different geographical locations were compared to recognize any variation in their venom composition. These traditional methods provided valuable information on the venom composition of cobras from different geographical locations of the country (Kalita & Mukherjee, 2019).

In 1998, Mukherjee and Maity from the Department of Biochemistry, Burdwan Medical College, West Bengal, studied the varied composition of *N. naja* venom samples from three districts (Burdwan, Purulia, and Midnapore) of West Bengal, eastern India, and western India (Haffkine Institute, Maharashtra) by liquid chromatographic separation of venom proteins (toxins) and the subsequent biochemical assays of crude cobra venom and each fraction of venom (Mukherjee and Maity, 1998). These venoms were decomplexed using size-exclusion (gel filtration) chromatography on Sephadex G-50 and by cation-exchange chromatography on a CM-Sephadex C-25 column. The venom fractions and crude venom were analyzed for protease activity on casein and plasma proteins, esterase, phospholipase A_2, and $5'$-nucleotidase enzymes. The results of the enzyme assays are shown in Table 4.5. The study was the first to demonstrate the disparity in the composition of Indian cobra (*N. naja*) venom from three adjoining districts of West Bengal, an eastern state of India. Further, the intrazonal variation in venom composition also results in different toxicities (lethality) and pharmacological properties of *N. naja* venom from eastern India (Mukherjee and Maity, 1998).

The above study also revealed that due to the variations in the composition of Indian cobra venom samples from different locales of India, the polyvalent antivenom raised against the *N. naja* venom from a particular geographical region may not be equally effective in neutralizing the venom components of *N. naja* from other locales of India (Mukherjee and Maity, 1998). Shashidharamurthy et al. (2002) also studied the variation in venom composition of *N. naja* from eastern (West Bengal), western (Maharashtra), and southern (Tamil Nadu) India (Shashidharamurthy et al., 2002, 2010). The SDS-PAGE analysis of the *N. naja* venom samples exhibited different banding patterns; the low-molecular-mass proteins predominated in eastern India cobra venom, in comparison to that from the other two regions (Shashidharamurthy et al., 2002). More than 10 years after this study, the findings have been substantiated by proteomic analysis of Indian cobra venoms in our laboratory (see below). Further, eastern India (*N. naja*) venom displayed higher

Table 4.5 Comparative analysis of the enzyme activity of *N. naja* venom samples obtained from eastern India (Burdwan, Purulia, Midnapore) and western India (Maharashtra)

Enzyme parameters (U/mg of protein)	Enzyme activity of venom samples			
	Eastern India		Western India	
	Burdwan	Purulia	Midnapore	Maharashtra
Protease activity (substrates)				
(a) Gelatin ($n = 3$)[1]	ND	ND	ND	ND
(b) Hide powder ($n = 3$)[2]	4.3 ± 0.08	4.0 ± 0.1	4.1 ± 0.1	4.1 ± 0.08
(c) Casein ($n = 6$)[3]	$3.1 \pm 0.18^{b,y}$	$3.2 \pm 0.2^{a,x}$	3.5 ± 0.22	3.6 ± 0.25
(d) Bovine serum albumin ($n = 4$)[3]	$0.48 \pm 0.07^{c,y}$	$0.41 \pm 0.05^{a,x}$	0.32 ± 0.08	0.24 ± 0.08
(e) Albumin ($n = 4$)[3]	$0.59 \pm 0.12^{b,y}$	0.46 ± 0.10^{a}	0.31 ± 0.10	0.24 ± 0.08
(f) Globulin ($n = 4$)[3]	0.47 ± 0.11^{a}	0.40 ± 0.08^{a}	0.3 ± 0.10	0.24 ± 0.09
(g) Fibrinogen ($n = 4$)[3]	$1.06 \pm 0.10^{b,y}$	$0.93 \pm 0.11^{a,x}$	0.69 ± 0.08	0.71 ± 0.10
Esterase activity (substrates)				
(a) BTEE ($n = 3$)[1]	ND	ND	ND	ND
(b) TAME ($n = 3$)[1]	ND	ND	ND	ND
Phospholipase A_2 ($n = 6$)[4]	$114 \pm 8.0^{c,z}$	$107 \pm 8.0^{b,x}$	95 ± 5.0	91 ± 6.0
5′-Nucleotide activity ($n = 6$)[5]	$133 \pm 10.0^{c,x}$	128 ± 9.0^{c}	118 ± 11.0^{c}	65 ± 6.0

Table adopted with permission from Mukherjee and Maity (1998)
ND Not detected. Values are mean $\pm$ SD of n number of observations.
Significantly different with respect to venom sample from Haffkine Institute: [a]$P < 0.05$, [b]$P < 0.01$, [c]$P < 0.001$. Significantly different compared to venom sample from Midnapore: [x]$P < 0.05$, [y]$P < 0.01$, [z]$P < 0.001$.
[1](+) represents proteolytic activity for a particular substrate whereas (−) represents no activity
[2]Specific activity was determined by dividing the optical density at 595 nm by mg of venom protein used
[3]Unit is defined as *n* mole equivalent of tyrosine formed per minute
[4]Unit is defined as *n* moles of fatty acid formed per minute
[5]Incubation was carried out at 37 °C for 30 min. Unit is defined as mg of protein

phospholipase A_2, hyaluronidase, and indirect hemolytic activities, in comparison to the venom from the same species of cobra from western and southern regions of the country. Since the low-molecular-mass 3FTxs of cobra venom cumulatively account for the toxicity of cobra venom (Kini & Doley, 2010), this would explain the decreased lethality (LD_{50}) of eastern (0.7 mg/kg), western (1.2 mg/kg), and southern (2.0 mg/kg) India *N. naja* venoms being associated with a decreased 3FTx content in these venoms (Shashidharamurthy et al., 2002, 2010).

Because of the problems associated with the conventional biochemical analysis of snake venom, in recent years, scientists from India have shown a profound interest in the proteomics analysis of snake venom proteomes including cobra venom proteomes (Kalita & Mukherjee, 2019; Mukherjee, 2020). The analysis of snake venom proteins by tandem mass spectrometry has played a greater role than conventional biochemical analysis in the fast, easy, and efficient identification and the accurate determination of relative abundance of snake venom toxins and their complexes (Kalita & Mukherjee, 2019). The venom proteome composition of Indian cobra (*N. naja*) from different parts of the Indian subcontinent has been studied by shotgun proteomics (Sintiprungrat et al., 2016; Chanda, Kalita, et al., 2018a; Chanda & Mukherjee, 2020a, Chanda & Mukherjee, 2020b; Dutta et al., 2017; Ali et al., 2013). Pooled *N. naja* venom samples from Sri Lanka (Sintiprungrat et al., 2016), Pakistan (Wong et al., 2018), eastern India (Dutta et al., 2017; Chanda, Patra, et al., 2018b), western India (Chanda, Kalita, et al., 2018a), and southern India (Choudhury et al., 2017; Chanda & Mukherjee, 2020b) have been analyzed by LC-MS/MS analysis and revealed qualitative and quantitative differences in their venom composition. The number of toxins in these venom samples varies from 28 to 81, which reinforces the geographical variation in venom composition of this species of snake (reviewed by Kalita & Mukherjee, 2019).

Tandem mass spectrometry analysis of *N. naja* venom samples from the Indian subcontinent has further demonstrated that irrespective of geographical origin, low-molecular-mass 3FTxs predominate in Indian cobra venom, though their proportions vary depending on the geographical locale of the snake (Sintiprungrat et al., 2016; Dutta et al., 2017; Wong et al., 2018; Chanda, Kalita, et al., 2018a). For example, proteomic analysis from our laboratory has shown that the 3FTxs content of *N. naja* venom from eastern, western, and southern India is 61.3, 52.9, and 437%, respectively (Dutta et al., 2017; Chanda, Kalita, et al., 2018a; Chanda & Mukherjee, 2020b). Further, this difference in venom composition results in differences in toxicity (LD_{50}) and pathophysiological manifestations exhibited by *N. naja*-envenomed patients in different regions of the country (Shashidharamurthy et al., 2002, 2010; Dutta et al., 2017; Wong et al., 2018; Chanda, Kalita, et al., 2018a).

A list of the relative abundance of toxins in the venom of the Indian cobra from different locales of the country by proteomic analysis is shown in Table 4.6.

4.6   Genomic and Transcriptomic Analyses of Indian Spectacled Cobra Venom Toxins

Very recently, Suryamohan et al. (2020) used modern genomic technologies to demonstrate a de novo near-chromosome-level reference genome assembly of Indian cobra (two snakes from Kerala state of southern India and four snakes from the Kentucky Reptile Zoo, USA) *N. naja* venom to understand the genomic organization of its toxin genes, genetic variability, evolution of venom toxin genes, and their expression in venom glands. Flow cytometry and cytogenetic analysis revealed the size of the Indian cobra genome at 1.48–1.77 Gb with its diploid karyotype ($2n = 38$). The Indian cobra venom genome encompasses a pair of sex chromosomes (ZZ male or ZW female), 7 pairs of macrochromosomes (MACs), and 11 pairs of microchromosomes (MICs) with an average GC content of 40.46% (Suryamohan et al., 2020). The authors expected the 23,248 protein-coding genes responsible for generating 31,447 transcripts to be comprised of alternatively spliced products encoding 31,036 predicted proteins.

Table 4.6 Comparison of the relative distribution of toxins in Indian cobra (*N. naja*) venom

Toxin class	Relative distribution (%)			
	EI (B[*,a])	EI (N[*,b])	WI[c]	SI[d]
Enzymatic proteins				
Phospholipase A$_2$	20.2	11.4	7.0	18.2
Phospholipase B	0.01		0.04	ND
Snake venom metalloprotease	6.1	1.0	6.6	16.2
Snake venom serine protease	ND	0.3	0.7	0.01
Acetylcholinesterase	0.2	6.3	0.1	0.32
Cholinesterase	0.2	6.0	ND	ND
Phosphodiesterase	0.6	2.1	0.8	ND
Nucleotidase	1.2	0.4	1.4	0.2
L-amino-acid oxidase	1.9	0.8	1.2	11.9
Aminopeptidase	ND	ND	0.1	ND
Nonenzymatic proteins				
Three-finger toxins	61.1	63.8	68.5	37.4
Cobra venom factor	1.7	1.1	0.6	4.3
Cysteine-rich secretory protein	3.03	2.1	2.9	4.7
Kunitz-type serine protease inhibitor	ND	0.4	4.1	1.4
Ohanin-like protein	0.5	1.3	3.8	1.6
Nerve growth factor	3.1	0.9	1.1	0.6
Cystatin	0.12	ND	0.4	0.3
Natriuretic peptides	ND	2.0	ND	2.9

Relative abundance was determined by label-free quantitative proteomic analysis. *B and *N represent Burdwan and Nadia districts, respectively, of West Bengal. Eastern India (EI); western India (WI); southern India (SI)
ND Not detected.
[a]Chanda, Patra, et al. (2018b)
[b]Dutta et al. (2017)
[c]Chanda, Kalita, et al. (2018a)
[d]Chanda and Mukherjee (2020b)

A total of 139 *N. naja* venom toxin genes belonging to 33 gene families have been identified and 16 major toxin gene families have been organized on MACs. Of the 139 genes, 19 that have shown venom gland-specific expression are deduced to be responsible for the expression of venom-specific toxins (VSTs). The identified genes are responsible for expressing 19 3FTxs, including 14 conventional 3FTxs, 8 SVMPs, and 6 cysteine (Cys)-rich secretory venom proteins (CRISPs). In addition, other toxin gene families have been identified for PLA$_2$, phospholipase B, hyaluronidases, LAAO, ohanin, cathelicidins, natriuretic peptide, 5′-nucleotidase, KSPI, CVF, VEGF, platelet-derived growth factor (PDGF), placenta growth factor (PGF), NGF, and cystatin. The 3FTxs class of cobra venom has also been shown to contain 7 neurotoxins (both short- and long-chain neurotoxins), 6 cytotoxins, 4 cardiotoxins, 1 anticoagulant, and 1 muscarinic toxin gene.

4.7   Species-Specific Differences in the Venom Composition Between *N. naja* and *N. kaouthia* from the Same Geographical Location of the Country

Indian monocled cobra or Bengal cobra (*N. kaouthia*) is another abundant species of cobra prevalent in West Bengal, eastern India, and northeastern India (Mukherjee & Maity, 2002; Mukherjee, 2020). This species is also widespread in neighboring countries, such as Bangladesh, Nepal, Myanmar, and south-western China (Whitaker et al., 2004). This is one of the lethal snakes of Thailand responsible for most of the snakebite deaths (Mukherjee & Maity, 2002; Kulkeaw et al., 2007). The detailed differences in venom composition between Indian *N. naja* and Indian *N. kaouthia* have been resolved by biochemical and proteomic studies from our laboratory, which also show the differences in some enzymatic activities and pharmacological properties between *N. kaouthia* and *N. naja* venom originating from eastern India (Mukherjee & Maity, 2002; Chanda, Patra, et al., 2018b; reviewed by Kalita & Mukherjee, 2019; Mukherjee, 2020).

A comparison of enzyme activities showed that eastern India *N. naja* venom contains less caseinolytic activity ($P < 0.001$), plasma protease activity ($P < 0.05$), and acetylcholinesterase activity ($P < 0.01$), when compared to *N. kaouthia* venom (Mukherjee & Maity, 2002). Further, *N. kaouthia* venom does not contain the L-amino acid oxidase enzyme, which is present in *N. naja* venom. These venom samples did not display significant differences ($p > 0.05$) in PLA$_2$, adenosine monophosphatase, and adenosine triphosphatase enzyme activities (Mukherjee & Maity, 2002). A comparative analysis of the enzyme activities in *N. naja* and *N. kaouthia* is shown in Table 4.7.

The two cobra venoms in Table 4.7 show different chromatographic elution profiles from the size-exclusion chromatography (Sephadex G-50) under identical experimental conditions. Each gel filtration fraction also showed different enzyme activities, and protein and carbohydrate contents (Mukherjee & Maity, 2002). The *N. naja* and *N. kaouthia* venom samples also showed differences in lethality and pharmacological effects on BALB/c mice (Table 4.8). The authors suggested that the presence of higher quantities of basic PLA$_2$ and low-molecular-mass polypeptides, mainly represented by the 3FTxs of cobra venom (*N. naja* venom compared to *N. kaouthia* venom), may account for the greater lethality of the former venom (Mukherjee & Maity, 2002).

Because commercial polyvalent antivenom contains equine antibodies only against Indian *N. naja* venom and not against Indian *N. kaouthia* venom, we also compared the potency of a commercial polyvalent antivenom for neutralizing the lethality and pathophysiology of *N. naja* and *N. kaouthia* venoms from eastern India (Mukherjee & Maity, 2002). We found that the polyvalent antivenom that contained *N. naja* antibodies poorly neutralized the various pathological effects of *N. kaouthia* envenomation. Thus, we concluded that antibodies against *N. kaouthia* venom, instead of just against *N. naja* venom, might be more beneficial for the treatment of *N. kaouthia* bite (Mukherjee & Maity, 2002). Furthermore, we proposed that Indian polyvalent antivenoms could also be supplied with antibodies against

Table 4.7 Comparison of enzyme activities of *N. naja* and *N. kaouthia* venoms from the Burdwan district of eastern India

Enzyme parameters	Enzyme activity (unit/mg protein)	
	N. naja	*N. kaouthia*
Proteolytic activities (substrates)		
Casein[a]	3.0 ± 0.17	5.2 ± 0.4*
Bovine serum albumin[a]	0.48 ± 0.07	0.48 ± 0.07
Human serum albumin[a]	0.59 ± 0.12	0.58 ± 0.11
Human serum globulin[a]	0.47 ± 0.11	0.46 ± 0.10
Human plasma fibrinogen[a]	1.06 ± 0.10	1.18 ± 0.12***
Phospholipase A_2[b]	116 ± 8.0	110 ± 7.0
Adenosine monophosphatase[c]	130 ± 8.0	118 ± 9.0
Adenosine triphosphatase[c]	86.5 ± 3.0	100 ± 5.5
Acetylcholinesterase[d]	17.0 ± 1.8	26 ± 1.5**
L-amino acid oxidase[e]	Present	Absent

Reprinted with permission from Mukherjee & Maity (2002)
Values are mean_S.D. of triplicate determinations.
[a]Unit is defined as nmole equivalent of tyrosine formed per minute
[b]Unit is defined as nmole of fatty acid formed per minute
[c]Unit is defined as μg Pi liberated per 30 min at 37 °C
[d]Micromoles of thiocholine formed per minute
[e]Solution of L-leucine (0.9%) in 0.05 M borate buffer pH (8.2) was used as substrate for qualitative assay of L-amino acid oxidase activity
*$P < 0.001$. Significance of difference (Student's *t*-test) as compared to *N. naja* venom
**$P < 0.01$. Significance of difference (Student's *t*-test) as compared to *N. naja* venom
***$P < 0.05$. Significance of difference (Student's *t*-test) as compared to *N. naja* venom

Table 4.8 Comparison of lethality (LD_{50}) and some pathophysiological effects of *N. naja* and *N. kaouthia* venoms on albino BALB/c mice

Properties	Biological activity	
	N. naja	*N. kaouthia*
Lethality (mg/kg body wt. of mouse)	0.40 ± 0.08	0.7 ± 0.09*
Hemorrhagic activity (20 μg crude venom)	Nil	Nil
Myotoxicity (CPK) release(IU/L)[a]	28.5 ± 3.0	19.2 ± 2.1
Edema ratio (15 μg crude venom)	148 ± 3.1	140 ± 3.1**
Minimum edema-inducing dose (μg)	5.0	4.8
Indirect hemolytic activity (% of Hb released by 100 μg protein)	86 ± 4.1	80 ± 3.8
Direct hemolytic activity (% of Hb released by 100 μg protein)	50 ± 2.1	39 ± 2.5**
Neurotoxicity	Present	Present

Reprinted with permission from Mukherjee and Maity (2002)
Values are mean ± S.D. of four determinations. Significance of difference (Student's *t*-test) compared to *N. naja* venom
*$P < 0.05$
**$P < 0.01$
[a]CPK value for control animals: 10.0 ± 1.3 IU/L

N. kaouthia venom to afford better protection against envenomation by this species of cobra (Chanda, Patra, et al., 2018b; Mukherjee, 2020).

4.8   Pharmacology, Pathophysiology, and Clinical Features of Indian Spectacled Cobra Envenomation

Because the quantity of venom injected into the prey or victim is under the voluntary control of the snake, depending on the size of the target and the size of the Indian cobra, approximately 170–250 mg of venom can be injected by *N. naja* into its victim. Envenomation by *N. naja* produces an array of pathophysiological and clinical symptoms that also depend on the age and size of the victim and if not treated immediately, it can lead to death or permanent morbidity (Britt & Burkhart, 1997; Mukherjee & Maity, 2002). Thus, the Indian cobra is classified as a category I medically important snake in India (World Health Organization, 1981).

Immediately after a cobra bite, the patient feels a burning and intense pain at the site of the bite and swelling usually occurs 2–3 h post-bite (Dass et al., 1998; Mukherjee, 2020; Mukherjee et al., 2021). In cases of a dry bite or when less venom is injected, swelling may not occur (Dass et al., 1998). The most prominent clinical symptom exhibited by *N. naja*-envenomed patients is neurotoxicity, which is characterized by a blurring of vision, diplopia, dysconjugate gaze, and ptosis (Britt & Burkhart, 1997; Kularatne et al., 2009; Mukherjee, 2020; Mukherjee et al., 2021). Cobra-envenomed patients have also been frequently reported to show drowsiness and irritability, which can be preceded by corticobulbar tract dysfunction such as dysphonia, dysphagia, absent gag reflex, and respiratory compromise (Britt & Burkhart, 1997; Kularatne et al., 2009). Finally, a patient's diaphragmatic breathing may lead to ventilator paralysis, respiratory deficiency, and then hypoxia and acidosis that requires immediate treatment; otherwise death usually occurs (Britt & Burkhart, 1997; Kularatne et al., 2009). A respiratory-compromised patient, who displays shallow and rapid respiration, may also show flaccid paralysis, convulsion, and vomiting (Dass et al., 1998; Mukherjee, 2020; Mukherjee et al., 2021). The mean onset of paralysis, which depends on the quantity of venom injected, and the age and size of the victim, can occur 30 min to 6 h post-bite (Britt & Burkhart, 1997; Dass et al., 1998; Mukherjee, 2020; Mukherjee et al., 2021). Paralysis usually begins from the lower limbs (Dass et al., 1998).

Cobra venom factors (CVFs) account for complement activation (Vogel et al., 1996; Vogel & Fritzinger, 2010), while vespryns and natriuretic peptides of cobra venom are responsible for the locomotor dysfunction and hypotensive effects, respectively (Pung et al., 2005; Zhang et al., 2011). The neurotoxicity of cobra venom has also been reported to be due to acetylcholinesterase and neurotoxic PLA_2s, which are postsynaptic neurotoxins that bind reversibly to nerve terminals to impair nerve signaling (Endo & Tamiya, 1987; Ranawaka et al., 2013; Bhat et al., 1991; Machiah & Gowda, 2006).

Indian cobra bite also causes several local effects such as swelling of the bitten limb, edema, and extensive necrosis (Britt & Burkhart, 1997; Kularatne et al., 2009; Mukherjee, 2020; Mukherjee et al., 2021). Transient coagulopathy, which is

Table 4.9 Cobra venom toxins responsible for the pharmacological activity and clinical symptoms of cobra envenomation

Clinical symptoms	Responsible toxins	References
Neurotoxic symptoms and respiratory failure	Neurotoxins, acetylcholinesterase, neurotoxic PLA_2s	Ranawaka et al. (2013), Bhat et al. (1991), Machiah and Gowda (2006), Endo and Tamiya (1987)
Cardiotoxicity	Cardiotoxins, cytotoxins	Dufton and Hider (1988)
Complement activation	CVFs	Vogel et al. (1996)
Locomotor dysfunction	Vespryns	Pung et al. (2005)
Hypotension	Natriuretic peptides	Zhang et al. (2011)
Edema and local swelling	PLA_2s, metalloproteases, and hyaluronidase	Girish, Shashidharamurthy, et al. (2004b), Ponnappa et al. (2008)
Necrosis	PLA_2s, metalloproteases, and hyaluronidase	Girish, Mohanakumari, et al. (2004a), Girish, Shashidharamurthy, et al. (2004b), Ponnappa et al. (2008)
Transient coagulopathy	PLA_2	Doley and Mukherjee (2003), Mukherjee et al. (2014), Dutta et al. (2015)
Membrane damage	PLA_2	Doley et al. (2004)

Table 4.10 Intrazonal variation in lethality, and hemolytic and edema-inducing activity of *N. naja* venom samples from three neighboring districts of eastern India

Parameters	Venom samples		
	Burdwan	Purulia	Midnapore
Lethality LD_{50} (mg/kg body weight of mice by subcutaneous injection)	0.40 ± 0.08	0.50 ± 0.08	0.63 ± 0.11
Edema ratio (%)	148.0 ± 3.10	138.0 ± 2.50	124 ± 2.30
Hemolysis (% of Hb release)	50 ± 2.10	46.0 ± 1.80	33.0 ± 2.00

Reproduced with permission from Mukherjee and Maity (1998)
Values are mean $\pm$ SD of four observations; *Hb* hemoglobin

characterized by an increase in the 20-min whole-blood clotting time (20WBCT), has also been reported in cobra-bite patients in Sri Lanka and western India (Kularatne et al., 2009; Bawaskar et al., 2008). A list of cobra venom toxins responsible for different clinical symptoms post-envenomation is presented in Table 4.9.

The geographic variation in the venom composition of *N. naja* is well documented. The variation also results in different toxicities and lethalities of *N. naja* venom samples from three neighboring regions of eastern India (Table 4.10). In vivo studies have shown that Indian cobra venom samples from eastern India possess higher lethality, edema-inducing activity, and hemolysis, in comparison to the properties of *N. naja* venom from western India (Maharashtra). The authors suggested that, as a consequence of the variation, the cobra-bite patients from different regions may also display different severities of pathogenesis and clinical symptoms (Mukherjee and Maity, 1998; Mukherjee, 2020).

In a nutshell, further studies are necessary to understand the intrazonal and interzonal variations in the venom composition of the Indian cobra (*N. naja*) and the effect that this variation has on the venom neutralization capacity of polyvalent antivenoms. Clinical studies would also help to understand the pathophysiological manifestations in Indian cobra-bite patients admitted to hospitals in different parts of the country. By recognizing how the variation in venom composition is linked to the pathophysiological effects, specific antivenom therapies can be tailored for the treatment of Indian cobra bite.

References

Ali, S. A., Yang, D. C., Jackson, T. N., Undheim, E. A., Koludarov, I., Wood, K., Jones, A., Hodgson, W. C., McCarthy, S., Ruder, T., & Fry, B. G. (2013). Venom proteomic characterization and relative antivenom neutralization of two medically important Pakistani elapid snakes (*Bungarus sindanus* and *Naja naja*). *Journal of Proteomics, 89*, 15–23.

Babu, A. S., Puri, K. D., & Gowda, T. V. (1995). Primary structure of a cytotoxin-like basic protein from *Naja naja naja* (Indian cobra) venom. *International Journal of Peptide and Protein Research, 46*(1), 69–72.

Basavarajappa, B. S., & Gowda, T. V. (1992). Comparative characterization of two toxic phospholipases A_2 from Indian cobra (*Naja naja naja*) venom. *Toxicon, 30*(10), 1227–1238.

Bawaskar, H. S., Bawaskar, P. H., Punde, D. P., Inamdar, M. K., Dongare, R. B., & Bhoite, R. R. (2008). Profile of snakebite envenoming in rural Maharashtra, India. *The Journal of the Association of Physicians of India, 56*, 88–95.

Bhat, M. K., & Gowda, T. V. (1989). Purification and characterization of a myotoxic phospholipase A_2 from Indian cobra (*Naja naja naja*) venom. *Toxicon, 27*(8), 861–873.

Bhat, M. K., & Gowda, T. V. (1991). Isolation and characterization of a lethal phospholipase A_2 (NN-IVb1-PLA$_2$) from the Indian cobra (*Naja naja naja*) venom. *Biochemistry International, 25*(6), 1023–1034.

Bhat, M. K., Prasad, B. N., & Gowda, T. V. (1991). Purification and characterization of a neurotoxic phospholipase A_2 from Indian cobra (*Naja naja naja*) venom. *Toxicon, 29*, 1345–1349.

Bieber, A. (1979). Metal and nonprotein constituents in snake venoms. In C. Y. Lee (Ed.), *Snake venoms* (pp. 295–306). Springer.

Bogert, C. M. (1943). Dentitional phenomena in cobras and other elapids, with notes on adaptive modifications of fangs. *Bulletin of the American Museum of Natural History, 81*, 3.

Britt, A., & Burkhart, K. (1997). *Naja naja* cobra bite. *The American Journal of Emergency Medicine, 15*(5), 529–531.

Chaim-Matyas, A., Borkow, G., & Ovadia, M. (1995). Synergism between cytotoxin P4 from the snake venom of *Naja nigricollis nigricollis* and various phospholipases. *Comparative Biochemistry and Physiology Part B: Biochemistry and Molecular Biology, 110*(1), 83–89.

Chanda, A., Kalita, B., Patra, A., Sandani, W. D., Senevirathne, T., & Mukherjee, A. K. (2018a). Proteomic analysis and antivenomics study of Western India *Naja naja* venom: Correlation between venom composition and clinical manifestations of cobra bite in this region. *Expert Review of Proteomics, 16*(2), 171–184.

Chanda, A., Patra, A., Kalita, B., & Mukherjee, A. K. (2018b). Proteomics analysis to compare the venom composition between *Naja naja* and *Naja kaouthia* from the same geographical location of eastern India: Correlation with pathophysiology of envenomation and immunological cross-reactivity towards commercial polyantivenom. *Expert Review of Proteomics, 15*(11), 949–961.

Chanda, A., & Mukherjee, A. K. (2020a). Mass spectrometry analysis to unravel the venom proteome composition of Indian snakes: Opening new avenues in clinical research. *Expert Review of Proteomics, 17*(5), 411–423.

Chanda, A., & Mukherjee, A. K. (2020b). Quantitative proteomics to reveal the composition of Southern India spectacled cobra (*Naja naja*) venom and its immunological cross-reactivity towards commercial antivenom. *International Journal of Biological Macromolecules, 160*, 224–232.

Charles, A. K., Gangal, S. V., & Joshi, A. P. (1981). Biochemical characterization of a toxin from Indian cobra (*Naja naja naja*) venom. *Toxicon, 19*, 295–303.

Choudhury, M., McCleary, R. J., Kesherwani, M., Kini, R. M., & Velmurugan, D. (2017). Comparison of proteomic profiles of the venoms of two of the 'Big Four' snakes of India, the Indian cobra (*Naja naja*) and the common krait (*Bungarus caeruleus*), and analyses of their toxins. *Toxicon, 135*, 33–42.

Das, T., Bhattacharya, S., Halder, B., Biswas, A., Das, G., Gomes, S., A., & Gomes, A. (2011). Cytotoxic and antioxidant property of a purified fraction (NN-32) of Indian *Naja naja* venom on Ehrlich ascites carcinoma in BALB/c mice. *Toxicon, 57*, 1065–1072.

Dass, B., Bhatia, R., & Singh, H. (1998). Venomous snakes in India and management of snakebite. In B. D. Sharma (Ed.), *Snakes in India: A source book* (pp. 257–268). Asiatic Publishing house.

Deufel, A., & Cundall, D. (2006). Functional plasticity of the venom delivery system in snakes with a focus on the poststrike prey release behavior. *Zoologischer Anzeiger – A Journal of Comparative Zoology, 245*(3–4), 249–267.

Devi, A. (1968). The protein and nonprotein constituents of snake venoms. In *Venomous animals and their venoms* (pp. 119–165). Elsevier.

Doley, R., & Mukherjee, A. K. (2003). Purification and characterization of an anticoagulant phospholipase A_2 from Indian monocled cobra (*Naja kaouthia*) venom. *Toxicon, 41*, 81–91.

Doley, R., King, G. F., & Mukherje, A. K. (2004). Differential hydrolysis of erythrocyte and mitochondrial membrane phospholipids by two phospholipase A_2 isoenzymes (NK-PLA$_2$-I and NK-PLA$_2$-II), from Indian monocled cobra *Naja kaouthia* venom. *Archives of Biochemistry and Biophysics, 425*, 1–13.

Dufton, M. J., & Hider, R. C. (1988). Structure and pharmacology of elapid cytotoxins. *Pharmacology & Therapeutics, 36*, 1–40.

Dutta, S., Gogoi, D., & Mukherjee, A. K. (2015). Anticoagulant mechanism and platelet deaggregation property of a non-cytotoxic, acidic phospholipase A_2 purified from Indian cobra (*Naja naja*) venom: Inhibition of anticoagulant activity by low molecular weight heparin. *Biochimie, 110*, 93–106.

Dutta, S., Chanda, A., Kalita, B., Islam, T., Patra, A., & Mukherjee, A. K. (2017). Proteomic analysis to unravel the complex venom proteome of eastern India *Naja naja*: Correlation of venom composition with its biochemical and pharmacological properties. *Journal of Proteomics, 156*, 29–39.

Dutta, S., Archana Sinha, A., Dasgupta, S., & Mukherjee, A. K. (2019). Binding of a *Naja naja* venom acidic phospholipase A_2 cognate complex to membrane-bound vimentin of rat L6 cells: Implications in cobra venom-induced cytotoxicity. *Biochimica et Biophysica Acta – Biomembranes, 1861*, 958–977.

Endo, T., & Tamiya, N. (1987). Current view on the structure-function relationship of postsynaptic neurotoxins from snake venoms. *Pharmacology & Therapeutics, 34*(3), 403–451.

Fry, B. G., Wuster, W., Kini, R. M., Brusic, V., Khan, A., Venkataraman, D., & Rooney, A. P. (2003). Molecular evolution and phylogeny of elapid snake venom three-finger toxins. *Journal of Molecular Evolution, 57*(1), 110–129.

Girish, K.S.,Mohanakumari, H.P., Nagaraju, S, . Vishwanath, B.S., K. Kemparaju, (2004a) Hyaluronidase and protease activities from Indian snake venoms: Neutralization by *Mimosa pudica* root extract. Fitoterapia 75, 378–380.

Girish, K. S., Shashidharamurthy, R., Nagaraju, S., Gowda, T. V., & Kemparaju, K. (2004b). Isolation and characterization of hyaluronidase a "spreading factor" from Indian cobra (*Naja naja*) venom. *Biochimie, 86*(3), 193–202.

Gorai, B., & Sivaraman, T. (2017). Delineating residues for haemolytic activities of snake venom cardiotoxin 1 from *Naja naja* as probed by molecular dynamics simulations and in vitro validations. *International Journal of Biological Macromolecules, 95*, 1022–1036.

Jagadeesha, D. K., Shashidharamurthy, R., Girish, K. S., & Kemparaju, K. (2002). A non-toxic anticoagulant metalloprotease: Purification and characterization from Indian cobra (*Naja naja naja*) venom. *Toxicon, 40*, 667–675.

Jayanthi, G. P., & Gowda, T. V. (1983). Purification of acidic phospholipases from Indian cobra (*Naja naja naja*) venom. *Journal of Chromatography, 281*, 393–396.

Kalita, B., & Mukherjee, A. K. (2019). Recent advances in snake venom proteomics research in India: A new horizon to decipher the geographical variation in venom proteome composition and exploration of candidate drug prototypes. *Journal of Proteins and Proteomics, 10*, 149–164.

Kaneda, N., Takechi, M., Sasaki, T., & Hayashi, K. (1984). Amino acid sequence of cytotoxin IIa isolated from the venom of the Indian cobra (*Naja naja*). *Biochemistry International, 9*, 603–610.

Kini, R. M., & Doley, R. (2010). Structure, function and evolution of three-finger toxins: Mini proteins with multiple targets. *Toxicon, 56*, 855–867.

Kochva, E. (1987). The origin of snakes and evolution of the venom apparatus. *Toxicon, 25*, 65–106.

Koh, D., Armugam, A., & Jeyaseelan, K. (2006). Snake venom components and their applications in biomedicine. *Cellular and Molecular Life Sciences, 63*(24), 3030–3041.

Kularatne, S. A., Budagoda, B. D., Gawarammana, I. B., & Kularatne, W. K. (2009). Epidemiology, clinical profile and management issues of cobra (*Naja naja*) bites in Sri Lanka: First authenticated case series. *Transactions of the Royal Society of Tropical Medicine and Hygiene, 103*(9), 924–930.

Kulkeaw, K., Chaicumpa, W., Sakolvaree, Y., Tongtawe, P., & Tapchaisri, P. (2007). Proteome and immunome of the venom of the Thai cobra, *Naja kaouthia*. *Toxicon, 49*, 1026–1041.

Kumar, M. S., Girish, K. S., Vishwanath, B. S., & Kemparaju, K. (2011). The metalloprotease, NN-PF3 from *Naja naja* venom inhibits platelet aggregation primarily by affecting α2β1 integrin. *Annals of Hematology, 90*(5), 569–577.

Laustsen, A. H. (2016). Toxin synergism in snake venoms. *Toxin Reviews, 35*(3–4), 165–170.

Linnaeus, C. (1758). *Systema naturae per regna tria naturae: Secundum classes, ordines, genera, species, cum characteribus, differentiis, synonymis, locis_ (in Latin)* (10th ed.). Laurentius Salvius.

Machiah, D. K., & Gowda, T. V. (2006). Purification of a post-synaptic neurotoxic phospholipase A_2 from *Naja naja* venom and its inhibition by a glycoprotein from *Withania somnifera*. *Biochimie, 88*(6), 701–710.

McCleary, R. J. R., & Kini, R. M. (2013). Non-enzymatic proteins from snake venoms: A gold mine of pharmacological tools and drug leads. *Toxicon, 62*, 56–74.

Mebs, D. (1985). *List of biologically active components from snake venoms*. University of Frankfurt.

Meier, J., & Stocker, K. F. (1995). Biology and distribution of venomous snakes of medical importance and the composition of snake venoms. In J. White (Ed.), *Handbook of clinical toxicology of animal venoms and poisons* (pp. 367–412). CRC Press.

Mukherjee, A. K. (1998). *In: Some biochemical properties of cobra and Russell's viper venom and their some biological effects on albino rats*. Burdwan University, Burdwan.

Mukherjee, A. K. (2008). Phospholipase A_2-interacting weak neurotoxins from venom of monocled cobra *Naja kaouthia* display cell specific cytotoxicity. *Toxicon, 51*, 1538–1543.

Mukherjee, A. K. (2010). Non-covalent interaction of phospholipase A_2 (PLA_2) and kaouthiotoxin (KTX) from venom of *Naja kaouthia* exhibits marked synergism to potentiate their cytotoxicity on target cells. *Journal of Venom Research, 1*, 37–42.

Mukherjee, A. K. (2020). Species-specific and geographical variation in venom composition of two major cobras in Indian subcontinent: Impact on polyvalent antivenom therapy. *Toxicon, 188*, 150–158.

Mukherjee, A. K., & Maity, C. R. (1998). Composition of *Naja naja* venom sample from three district of West Bengal, Eastern India. *Comparative Biochemistry and Physiology Part A: Molecular & Integrative Physiology, 119*, 621–627.

Mukherjee, A. K., & Maity, C. R. (2002). Biochemical composition, lethality and pathophysiology of venom from two cobras--*Naja naja* and *N. kaouthia. Comparative Biochemistry and Physiology Part B, Biochemistry & Molecular Biology, 131*, 125–132.

Mukherjee, A. K., Kalita, B., & Thakur, R. (2014). Two acidic, anticoagulant PLA$_2$ isoenzymes purified from the venom of monocled cobra *Naja kaouthia* exhibit different potency to inhibit thrombin and factor Xa via phospholipids independent, non-enzymatic mechanism. *PLoS One, 9*(8), e101334.

Mukherjee, A. K., Kalita, B., Dutta, S., Patra, A., Maity, C. R., & Punde, D. (2021). Snake envenomation: Therapy and challenges in India. In S. P. Mackessy (Ed.), *Section V: Global approaches to envenomation and treatments, handbook of venoms and toxins of reptiles* (2nd ed.). CRC Press.

Nakai, K., Sasaki, T., & Hayashi, K. (1971). Amino acid sequence of toxin A from the venom of the Indian cobra (*Naja naja*). *Biochemical and Biophysical Research Communications, 44*, 893–897.

Neema, K. N., Vivek, H. K., Kumar, J. R., Priya, B. S., & Nanduja Swamy, S. (2015). Purification and biochemical characterization of L-amino acid oxidase from western region Indian cobra (*Naja naja*) venom. *International Journal of Pharmacy and Pharmaceutical Sciences, 7*, 167–171.

Ohta, M., Sasaki, T., & Hayashi, K. (1976). The primary structure of toxin B from the venom of the Indian cobra *Naja naja. FEBS Letters, 15*, 161–166.

Ohta, M., Sasaki, T., & Hayashi, K. (1981a). The amino acid sequence of toxin D isolated from the venom of Indian cobra (*Naja naja*). *Biochimica et Biophysica Acta (BBA) – Protein Structure, 671*, 123–128.

Ohta, M., Sasaki, T., & Hayashi, K. (1981b). Primary structure of toxin C from the venom of the Indian cobra (*Naja naja*). *Chemical & Pharmaceutical Bulletin, 29*, 1458–1475.

Ponnappa, K. C., Saviour, P., Ramachandra, N. B., Kini, R. M., & Gowda, T. V. (2008). INN-toxin, a highly lethal peptide from the venom of Indian cobra (*Naja naja*) venom – Isolation, characterization and pharmacological actions. *Peptides, 29*(11), 1893–1900.

Pung, Y. F., Wong, P. T. H., Kumar, P. P., Hodgson, W. C., & Kini, R. M. (2005). Ohanin, a novel protein from king cobra venom induces hypolocomotion and hyperalgesia in mice. *The Journal of Biological Chemistry, 280*, 13137–13147.

Ranawaka, U. K., Lalloo, D. G., & de Silva, H. J. (2013). Neurotoxicity in snakebite—The limits of our knowledge. *PLOS Neglected Tropical Diseases, 7*(10), e2302.

Rudrammaji, L. M., & Gowda, T. V. (1998). Isolation and characterization of an endogenous inhibitor of phospholipase A$_2$ from Indian cobra (*Naja naja naja*) venom. *Toxicon, 36*(4), 639–644.

Sachidananda, M. K., Murari, S. K., & Channe Gowda, D. (2007). Characterization of an antibacterial peptide from Indian cobra (*Naja naja*) venom. *Journal of Venomous Animals and Toxins including Tropical Diseases, 13*, 446–461.

Satish, S., Tejaswini, J., Krishnakantha, T. P., & Gowda, T. V. (2004). Purification of a class B1 platelet aggregation inhibitor phospholipase A$_2$ from Indian cobra (*Naja naja*) venom. *Biochimie, 86*(3), 203–210.

Shafqat, J., Beg, O. U., Yin, S.-J., & Zaidi, Z. H. (1990). Primary structure and functional properties of cobra (*Naja naja naja*) venom Kunitz-type trypsin inhibitor. *European Journal of Biochemistry, 194*, 337–341.

Shafqat, J., Siddiqi, A. R., Zaidi, Z. H., & Jörnvall, H. (1991). Extensive multiplicity of the miscellaneous type of neurotoxins from the venom of the cobra *Naja naja naja* and structural characterization of major components. *FEBS Letters, 284*, 70–72.

Shashidharamurthy, R., Jagadeesha, D. K., Girish, K. S., & Kemparaju, K. (2002). Variations in biochemical and pharmacological properties of Indian cobra (*Naja naja naja*) venom due to geographical distribution. *Molecular and Cellular Biochemistry, 229*(1–2), 93–101.

Shashidharamurthy, R., Mahadeswaraswamy, Y. H., Ragupathi, L., Vishwanath, B. S., & Kemparaju, K. (2010). Systemic pathological effects induced by cobra (*Naja naja*) venom from geographically distinct origins of Indian peninsula. *Experimental and Toxicologic Pathology, 62*(6), 587–592.

Sintiprungrat, K., Watcharatanyatip, K., Senevirathne, W. D., Chaisuriya, P., Chokchaichamnankit, D., Srisomsap, C., & Ratanabanangkoon, K. A. (2016). Comparative study of venomics of *Naja naja* from India and Sri Lanka, clinical manifestations and antivenomics of an Indian polyspecific antivenom. *Journal of Proteomics, 132*, 131–143.

Sudarsan, S., & Dhananjaya, B. L. (2015). Antimicrobial activity of an acidic phospholipase A_2 (NN-XIa-PLA2) from the venom of *Naja naja naja* (Indian cobra). *Applied Biochemistry and Biotechnology, 176*(7), 2027–2038.

Suryamohan, K., Krishnankutty, S. P., Guillory, J., et al. (2020). The Indian cobra reference genome and transcriptome enables comprehensive identification of venom toxins. *Nature Genetics, 52*, 106–117.

Suzuki-Matsubara, M., Athauda, S. B. P., Suzuki, Y., Matsubara, K., & Moriyama, A. (2016). Comparison of the primary structures, cytotoxicities, and affinities to phospholipids of five kinds of cytotoxins from the venom of Indian cobra, *Naja naja*. *Comparative Biochemistry and Physiology Part C: Toxicology & Pharmacology, 179*, 158–164.

Suzuki, M., Itoh, T., Anuruddhe, B. M., Bandaranayake, I. K., Shirani Ranasinghe, J. G., Athauda, S. B., & Moriyama, A. (2010). Molecular diversity in venom proteins of the Russell's viper (*Daboia russelii russelii*) and the Indian cobra (*Naja naja*) in Sri Lanka. *Biomedical Research, 31*, 71–81.

Takechi, M., Tanaka, Y., & Hayashi, K. (1987). Amino acid sequence of a less-cytotoxic basic polypeptide (LCBP) isolated from the venom of the Indian cobra (*Naja naja*). *Biochemistry International, 14*, 145–152.

Thangam, R., Gunasekaran, P., Kaveri, K., Sridevi, G., Sundarraj, S., Paulpandi, M., & Kannan, S. (2012). A novel disintegrin protein from *Naja naja* venom induces cytotoxicity and apoptosis in human cancer cell lines in vitro. *Process Biochemistry, 47*(8), 1243–1249.

Weinstein, S. A., Smith, T. L., & Kardong, K. V. (2009). Reptile venom glands form, function, and future. In *Handbook of venoms and toxins of reptiles* (pp. 65–91). CRC Press.

Suvilesh, K. N., Yariswamy, M., Savitha, M. N., Joshi, V., Nanjaraj, U. A. N., Urs, A. P., Choudhury, M., Velmurugan, D., & Vishwanath, B. S. (2017). Purification and characterization of an anti-hemorrhagic protein from *Naja naja* (Indian cobra) venom. *Toxicon, 140*, 83–93.

Vijayaraghavan, B. (2008). *Snakebite: A book for India* (pp. 1–93). The Chennai Snake Park Trust.

Vogel, C. W., & Fritzinger, D. C. (2010). Cobra venom factor: Structure, function, and humanization for therapeutic complement depletion. *Toxicon, 56*, 1198–1222.

Vogel, C. W., Bredehorst, R., Fritzinger, D. C., Grunwald, T., Ziegelmuller, P., & Kock, M. A. (1996). Structure and function of cobra venom factor, the complement-activating protein in cobra venom. *Advances in Experimental Medicine and Biology, 391*, 97–114.

Whitaker, R., Captain, A., & Ahmed, F. (2004). *Snakes of India*. Draco Books.

Whitaker, R. (2006). *Common Indian snakes: A field guide*. Macmillan Indian Pvt. Ltd..

Wong, K. Y., Tan, C. H., Tan, K. Y., Quraishi, N. H., & Tan, N. H. (2018). Elucidating the biogeographical variation of the venom of *Naja naja* (spectacled cobra) from Pakistan through a venom-decomplexing proteomic study. *Journal of Proteomics, 175*, 156–173.

World Health Organization. (1981). Progress in the characterization of venoms and standardization of antivenoms W.H.O. *WHO Offset Publication, 58*, 1–44.

Wüster, W. (1998). The cobras of the genus *Naja* in India. *Hamadryad, 23*(1), 15–32.

Wüster, W., & Thorpe, R. S. (1989). Population affinities of the Asiatic cobra (*Naja naja*) species complex in south-east Asia: Reliability and random resampling. *Biological Journal of the Linnean Society, 36*, 391–409.

Wüster, W., & Thorpe, R. S. (1992). Dentitional phenomena in cobras revisited: Spitting and fang structure in the Asiatic species of *Naja* (Serpentes: Elapidae). *Herpetologica, 48*(4), 424–434.

Zhang, Y., Wu, J., Yu, G., Chen, Z., Zhou, X., Zhu, S., Li, R., Zhang, Y., & Lu, Q. (2011). A novel natriuretic peptide from the cobra venom. *Toxicon, 57*(1), 134–140.

Indian Common Krait (*Bungarus caeruleus*)

5

Abstract

The Indian common krait (*Bungarus caeruleus*) or blue krait is one of the members of the "Big Four" venomous snakes of India. Black or bluish-black in color, it has an average length of 1–1.2 m and possesses a flat, blunt, short head with small eyes, with a neck that is barely visible. The snake has a distinctive feature of yellow and black cross-bands along its backbone. The common krait is distributed in the Indian subcontinent, from Afghanistan, the Sindh province of Pakistan, to the plains of West Bengal in eastern India, Bangladesh, and Nepal, though another species of krait (*Bungarus fasciatus*, the banded krait) predominates in the northeastern part of India. This snake is nocturnal (night dweller) and therefore the maximum number of krait bites are reported during late night. Although the venom is very potent, because of the poor yields, the biochemical composition of Indian common krait venom has not been analyzed in great detail. Only a few toxins from *B. caeruleus* venom have been purified and characterized though recent proteomic analyses have provided more comprehensive data on the occurrence of different toxins in its venom. The venom contains 57 distinct proteins distributed in 12 snake venom toxin families. The proteomic analyses indicate that PLA_2 (37.6%) and three-finger toxins (48.3%) are the most abundant enzymatic and nonenzymatic protein families, respectively. A comparison of two independent studies on the proteomic analyses of krait venom from two distinct geographical locales (southern India and Sri Lanka) has shown the differences in venom composition for *B. caeruleus* in these countries. Similar to Indian cobra bite, Indian krait bite patients also have neuroparalytic symptoms and also frequently experience abdominal pain and cramps due to internal bleeding. Further studies are required to characterize and compare the venom composition of the Indian common krait from different regions.

© The Author(s), under exclusive license to Springer Nature Singapore Pte Ltd. 2021
A. K. Mukherjee, *The 'Big Four' Snakes of India*,
https://doi.org/10.1007/978-981-16-2896-2_5

Keywords

Bungarus caeruleus · Clinical features of Indian common krait envenomation ·
Indian common krait · Pharmacology of *Bungarus caeruleus* venom ·
Neurotoxicity · Proteomic analysis of venom · Venom composition of the Indian
common krait

5.1   Taxonomic Classification of the Indian Common Krait (*Bungarus caeruleus*)

Phylum: Chordata
Group: Vertebrata
Subphylum: Gnathostomata
Class: Reptilia
Subclass: Diapsida
Order: Squamata
Suborder: Ophidia
Infraorder: Xenophidia
Family: Elapidae
Subfamily: Viperinae
Genus: *Bungarus*
Species: *caeruleus*

5.2   Characteristic Features of the Indian Common Krait

The Indian common krait (*Bungarus caeruleus*) or blue krait is regarded as one of
the most deadly snakes of the Indian subcontinent and Southeast Asia (Fig. 5.1).

The binomial name *Bungarus caeruleus* was coined by a German naturalist
Johann Gottlob Theaenus Schneider in 1801. The bite of this species of snake
requires immediate treatment, and therefore, the common krait is considered as
one of the "Big Four" venomous snakes of India. The Indian common krait is
known by several names in different Indian languages (Table 5.1).

On average, the Indian common krait can reach lengths of 1–1.2 m (3.3 ft), but
can also grow to 1.75 m (5 ft. 9 in). Males are longer than females and both possess a
flat, blunt, short head with small eyes, and have a neck that is barely visible
(Vijayaraghavan, 2008). The cylindrical body narrows toward the short and rounded
tail. Because of this unique shape, it is sometimes difficult to distinguish the mouth
from the tail (Khaire, 2014). The triangular body shape of the Indian common krait
helps it to move about in marshy environments and wetlands (Khaire, 2014). The
bright, shiny, and smooth scales occur in 15–17 rows and the vertebral row is
markedly distended and hexagonal in shape. The fangs are short and fixed in the
front part of the upper jaw beneath the nostrils.

Fig. 5.1 Photograph of the Indian common krait (*Bungarus caeruleus*) (PC Vivek Sharma)

Table 5.1 Vernacular names of the Indian common krait

Language	Local name
Bengali	কালাচ (kālāch), ডং মিনাচিতি (ḍōmnāciti)
Hindi	करैत (kareit)
Gujarati	કાળોતરો (kāḷōtarō)
Kannada	ಕಟ್ಟಿಗೆ ಹಾವು (kaṭṭigĕ hāvu)
Malayalam	വെള്ളിക്കെട്ടൻ (veḷḷikkeṭṭan)
Marathi	मणयार (maṇyār)
Odia	କଳେଣ୍ଟ (kalent)
Tamil	கட்டு விரியன் (kaṭṭu viriyaṇ)
Telegu	కట్ల పాము (kaṭla pāmu)

This species does not show much variation in color. They may be black or bluish-black and have approximately 40 thin, distinct white crossbars or -bands (in pairs or in a single row) at the posterior (tail region) of the body (Sharma, 1998a, 1998b, 1998c). Anterior crossbars are only seen in young snakes. The upper lips and belly are white.

5.3 Geographical Distribution, Habitat, Behavior, and Reproduction of the Indian Common Krait

The common krait is distributed in the Indian subcontinent, from Afghanistan, the Sindh province of Pakistan, to the plains of West Bengal in eastern India, Bangladesh, and Nepal. This species is also found in the southern part of the country and in Sri Lanka. It is found up to 1600 m (5250 ft) above sea level. In the northeastern part of India, two other species of krait (the banded krait, *B. fasciatus*, and the black krait, *Bungarus niger*) are prevalent (Fig. 5.2a, b). The distinctive appearance of the former species of krait is yellow and black cross-bands along its backbone (Sharma, 1998c), while the latter species of krait is shiny and black in color.

This snake can live in a wide variety of habitats including agricultural fields, bushy vegetation, and nearby villages. It prefers to reside in termite mounds, brick piles, old temples, mammalian holes, and farmlands (Vijayaraghavan, 2008). They can also be found near ponds and shallow bodies of water. Typically, these kraits

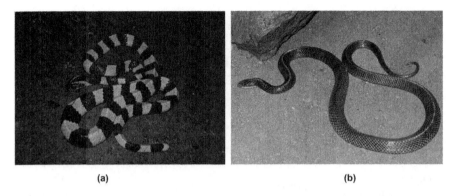

(a)   (b)

Fig. 5.2 (**a**) Indian banded krait (*Bungarus fasciatus*), (**b**) Indian black krait (*Bungarus niger*) (PC Jayaditya Purakaystha)

will enter human dwellings in search of food, such as rats, mice, and lizards (Khaire, 2014). These snakes can climb to the upper stories of houses in search of food (personal observation) and they will also eat frogs and rabbits that are found near agricultural fields. They can also eat other kraits (Khaire, 2014) and the young snakes may eat arthropods.

As an exclusively nocturnal species, this snake is not active during the day, when they appear to be quite submissive, sluggish, and reluctant to show any aggressive behavior. As a nocturnal species, they are active from late evening to early morning and tend to move quickly, displaying some aggressiveness and making loud hissing sounds when threatened. These snakes will bite humans only when they have no way to escape. Most cases of krait bite are reported during the late night when the snake enters a hut or house in search of food and it accidently bites a person who may be sleeping on the floor. The bite of the common krait is painless and the victim may not realize that they have been bitten. If a sufficiently large amount of venom is injected, the envenomed person may die while sleeping.

The breeding season of the Indian common krait is during the summer (from April to July/August). The female is oviparous and lays about a dozen eggs in holes or mounds. The eggs hatch after 2 months of incubation, while the female guards the eggs from predators.

5.4   Venom Composition of the Indian Common Krait

Unlike venom of the Indian cobra and Indian Russell's viper, the biochemical composition of Indian common krait venom has not been analyzed in great detail. As an explanation for this, the yield of venom from this snake is small (about 8–12 mg per milking), and therefore, it is costly and difficult to obtain. Nevertheless, common krait venom is extremely potent and 2–3 mg is lethal for humans (Murthy, 1998). The venom is neurotoxic and when injected, victims show neuroparalytic symptoms (Mukherjee et al., 2021).

Table 5.2   List of toxins purified and characterized from Indian common krait venom

Enzyme class	Name of toxin	Molecular weight (kDa)	Biological functions	References
Hyaluronidase	HYL	14 ± 2	Not described	Bhavya et al. (2016)
Neurotoxin	Neurotoxin peptides	Not mentioned	The purified neurotoxin showed moderate toxicity in treated mice, and demonstrated prominent local and systemic effects	Mirajkar et al. (2005)
	Beta 1-bungarotoxin (this heterodimeric neurotoxin consists of a PLA$_2$ subunit linked by a disulfide bond to a K+ channel-binding subunit, which is a member of the Kunitz protease inhibitor subfamily)	Not mentioned	Not mentioned	Sharma et al. (1999)

Only a few toxins from Indian common krait venom have been purified and characterized (Table 5.2) and the enzymatic activity of Indian common krait venom from southern India (SI) is shown in Table 5.3.

Sodium dodecyl sulfate-polyacrylamide gel electrophoresis (SDS-PAGE) analysis showed that more than 85% of the proteins of SI *B. caeruleus* venom are in the mass range of 5–15 kDa, suggesting that phospholipase A2 (PLA$_2$) PLA$_2$, three-finger toxins (3FTxs), and Kunitz-type serine protease inhibitors (KSPI) are the predominant toxins in this venom (Patra et al., 2019). The proteomic analysis provided comprehensive data on the occurrence of different toxins in Indian common krait venom. Choudhury et al. (2017) used proteomics to identify the toxins of *B. caeruleus* venom of southern India and reported the occurrence of several PLA$_2$ enzymes and nonenzymatic 3FTxs. They also identified nonenzymatic toxins like the cysteine-rich secretory proteins, venom nerve growth factors, and some enzymatic proteins such as metalloproteinases, L-amino acid oxidases, 5′-nucleotidases, acetylcholinesterase, and hyaluronidases in SI *B. caeruleus* venom (Choudhury et al., 2017).

The quantitative proteome composition of SI common krait venom was also analyzed in our laboratory (Patra et al., 2019). The venom was found to contain 57 distinct proteins distributed in 12 snake venom toxin families. The proteomic analysis showed that PLA$_2$ (37.6%) and 3FTxs (48.3%) are the most copious enzymatic and nonenzymatic protein families, respectively, in this venom. The 3FTxs of this venom are comprised of long-chain neurotoxin (LNTX) (20.0%),

Table 5.3 Enzymatic
activities of SI *B. caeruleus*
venom

Enzyme activities	Specific activity (U/mg)
ATPase[a]	$0.19 \pm 0.01 \times 10^3$
ADPase[a]	$0.05 \pm 0.002 \times 10^3$
AMPase[a]	$1.23 \pm 0.05 \times 10^3$
Hyaluronidase[b]	$0.70 \pm 0.03 \times 10^3$
PLA$_2$[c]	$0.76 \pm 0.03 \times 10^3$
SVMP[d]	0.01 ± 0.001
LAAO[e]	50.0 ± 1.90
Fibrinogenolytic activity	ND
Fibrinolytic activity	ND
PDE	ND
BAEE	ND
TAME	ND

Values are mean ± SD of triplicate determinations
Reprinted with permission from Patra et al. (2019)
[a]One unit is defined as micromoles of Pi released per minute
[b]One unit is defined as micromoles of p-nitrophenol released per
minute. [b]One unit is defined as a 1% decrease in turbidity at 405 nm
compared to the control (100% turbidity)
[c]One unit is defined as a decrease by 0.01 in absorbance at 740 nm
after 10 min of incubation
[d]One unit is defined as the change in absorbance at 450 nm per
minute at 37 °C
[e]One unit is defined as 1 nmol of kynurenic acid produced per minute
ND not detected
ATPase adenosine triphosphatase, *ADPase* adenosine
diphosphatase, *AMPase* adenosine monophosphatase, *PLA2* phospholipase A2, *SMVP* snake venom metalloprotease, *LAAO* L-amino
acid oxidase, *PDE* phosphodiesterase, *BAEE* Nα-Benzoyl-L-arginine ethyl ester, *TAME* N-alpha-tosyl-L-arginine methyl este

short-chain neurotoxins (SNTX) (0.7%), cardiotoxin (0.2%), and cobrotoxin (0.1%).
In addition, minor quantities (<5% relative distribution) of other toxins such as
SVMP (4.8%), cysteine-rich secretory proteins (CRISP, 5.0%), L-amino acid oxidase (LAAO, 2.1%), acetylcholinesterase (AchE, 1.3%), vespryn (0.6%), cobra
venom factor (CVF, 0.4%), snake venom serine protease (SVSP, 0.1%), phospholipase B (PLB, 0.1%), and Kunitz-type serine protease inhibitor (KSPI, 0.1%) were
identified in this venom (Patra et al., 2019). The LC-MS/MS data is well correlated
with the percent protein band intensity of the respective groups of toxins determined
by sodium dodecyl polyacrylamide gel electrophoresis (SDS-PAGE) analysis.

The venom of *B. caeruleus* from Sri Lanka (SL) was also studied by LC-MS/MS
analysis and found to contain the highest amount of phospholipases A_2 enzymes
(64.5% of total proteins) including 4.6% that are presynaptically acting
β-bungarotoxin A-chains (Oh et al., 2017). The nonenzymatic three-finger toxins
(19.0%), comprising 15.6% of the κ-neurotoxins (the potent postsynaptically acting

long neurotoxins), were the second most abundant toxin class in this venom. Based on their identical chromatographic elution profiles, the compositions of venoms from the Indian common krait of SL, India, and Pakistan were found to be similar (Oh et al., 2017). Still, in a comparison of two independent studies on the proteomic analysis of krait venom from two distinct sources, some differences were found in the composition of SI and SL *B. caeruleus* venom (Oh et al., 2017; Patra et al., 2019). Geographical differences in venom composition may be especially relevant for effective antivenom treatments since SL imports antivenom from India, which is produced against Indian *B. caeruleus* venom. Thus, SL *B. caeruleus* antivenom that is venom specific must be produced to improve the efficacy of snakebite treatment (Keyler et al., 2013).

5.5   Pharmacology, Pathophysiology, and Clinical Features of the Indian Common Krait Envenomation

The Indian common krait produces significantly less venom (approximately 10–12 mg) than that produced by the Indian cobra (see Chap. 4). Nevertheless, the potency of the former is much greater than that of the latter (Murthy, 1998). The primary clinical manifestations of envenomation by the Indian cobra and Indian common krait are similar neuroparalytic symptoms (blurring of vision, slurring of speech, difficulty in swallowing, drowsiness, and thickened tongue); however, common krait envenomation does not produce local symptoms and the bite is painless (Dass et al., 1998; Vijayaraghavan, 2008; Kularatne, 2002; Ariaratnam et al., 2008; Law et al., 2014; Mukherjee et al., 2021). Still, krait-bite patients frequently experience abdominal pain and cramps due to internal bleeding, which is followed by a tightening of facial muscles and respiratory paralysis within 6–24 h post-bite (Dass et al., 1998). If not treated immediately, krait bite may lead to death. Other clinical features of Indian common krait envenomation are nausea, vomiting, hypotension, shock, comma, and drowsiness (Dass et al., 1998; Ariaratnam et al., 2008). Hyperkalemia, an elevated potassium level in the blood, is also observed in Indian common krait-bite patients (Bawaskar & Bawaskar, 2015). A clear correlation between the occurrence of different toxins in Indian common krait venom and the clinical manifestations from common krait bite is presented by Patra et al. (2019). Table 5.4 shows the common krait venom toxins responsible for different clinical symptoms post-envenomation.

The in vivo pharmacological activity of common krait venom was also assessed in an animal model. A sublethal dose of *B. caeruleus* venom (25 µg/kg) was injected intramuscularly into rats and the animals were observed for 3 h postinjection (Kiran et al., 2004). Hyperglycemia (increase in blood sugar) and an increase in serum alkaline phosphatase and urea were observed in the treated group though the concentrations of serum cholesterol and triglycerides were unaltered. Histopathological examination of rat tissues revealed gross morphological alterations in the heart, kidney, and liver tissues. Hemorrhage, myocardial fiber necrosis, constriction of blood vessels in the kidney with congested vessels, necrosis of the proximal

Table 5.4 Indian common krait venom toxins responsible for pharmacological activity and clinical symptoms of cobra envenomation

Clinical symptoms	Responsible toxins	References
Neurotoxic symptoms (blurring of vision, loss of consciousness, and neuroparalysis)	Postsynaptic neurotoxins (short neurotoxins, long-chain neurotoxin, and cytotoxins), acetylcholinesterase	Ariaratnam et al. (2008), Schetinger et al. (2009)
Tightening of neck muscle and respiratory failure	Presynaptic neurotoxin (β-bungarotoxins)	Dixon and Harris (1999), Prasarnpun et al. (2004)
Local swelling and abdominal pain	SVMP and PLA_2 (other than β-bungarotoxins)	Bawaskar and Bawaskar (2015)
Hyperkalemia	Cysteine-rich secretory proteins (CRISPs) and Kunitz-type serine protease inhibitor (KSPI)	Bawaskar and Bawaskar (2015)
Myotoxicity and myocardial infraction	Synergistic actions of PLA_2, cytotoxins, and cardiotoxins	Punde (2005), Bawaskar and Bawaskar (2015)

tubules, and liver congestion appeared to be induced by the venom, suggesting that it damages internal organs (Kiran et al., 2004).

In conclusion, the Indian common krait is one of the deadliest snakes in the country and a member of the "Big Four" venomous snakes, though it is a less studied species. The venom of *B. caeruleus* produces neurotoxic symptoms including coma and hypertension that are unusual neurological complications of Indian common krait envenomation. Further clinical research is needed to understand the pathophysiological features of Indian common krait envenomation and its correlation to the venom composition of kraits in that region. In addition, research data on the efficacy of polyantivenom for treating krait envenomation would be valuable for toxinologists and antivenom manufacturers who are designing region-specific antivenoms. The purification and characterization of the potent toxins from this venom would help in evaluating their pharmacological activity, neutralizing their adverse effects by commercial antivenoms, and developing their therapeutic applications.

References

Ariaratnam, C. A., Sheriff, M. R., Theakston, R. D. G., & Warrell, D. A. (2008). Distinctive epidemiologic and clinical features of common krait (Bungarus caeruleus) bites in Sri Lanka. *The American Journal of Tropical Medicine and Hygiene, 79*, 458–462.

Bawaskar, H. S., & Bawaskar, P. H. (2015). Snake bite poisoning. *Journal of Mahatma Gandhi Institute of Medical Sciences, 20*, 5–14.

Bhavya, J., Vineetha, M. S., Sundaram, P. M., Veena, S. M., Dhananjaya, B. L., & More, S. S. (2016). Low-molecular weight hyaluronidase from the venom of Bungarus caeruleus (Indian common krait) snake: Isolation and partial characterization. *Journal of Liquid Chromatography & Related Technologies, 39*, 203–208.

Choudhury, M., McCleary, R. J., Kesherwani, M., Kini, R. M., & Velmurugan, D. (2017). Comparison of proteomic profiles of the venoms of two of the 'Big Four' snakes of India, the Indian cobra (Naja naja) and the common krait (Bungarus caeruleus), and analyses of their toxins. *Toxicon, 135*, 33–42.

Dixon, R. W., & Harris, J. B. (1999). Nerve terminal damage by β-bungarotoxin: Its clinical significance. *The American Journal of Pathology, 154*, 447–455.

Dass, B., Bhatia, R., & Singh, H. (1998). Venomous snakes in India and management of snakebite. In B. D. Sharma (Ed.), *Snakes in India: A source book* (pp. 257–268). Asiatic Publishing house.

Keyler, D. E., Gawarammana, I., Gutiérrez, J. M., Sellahewa, K. H., McWhorter, K., & Malleappa, H. R. (2013). Antivenom for snakebite envenoming in Sri Lanka: The need for geographically specific antivenom and improved efficacy. *Toxicon, 69*, 90–97.

Khaire, N. (2014). *Indian snakes: A field guide.* Jyotsna Prakasan.

Kiran, K. M., More, S. S., & Gadag, J. R. (2004). Biochemical and clinicopathological changes induced by Bungarus coeruleus venom in a rat model. *The Journal of Basic and Clinical Physiology and Pharmacology, 15*, 277–287.

Kularatne, S. (2002). Common krait (Bungarus caeruleus) bite in Anuradhapura, Sri Lanka: A prospective clinical study, 1996–98. *Postgraduate Medical Journal, 78*, 276–280.

Law, A. D., Agrawal, A. K., & Bhalla, A. (2014). Indian common krait envenomation presenting as coma and hypertension: A case report and literature review. *The Journal of Emergencies, Trauma, and Shock, 7*(2), 126–128.

Mirajkar, K. K., More, S., & Gadag, J. R. (2005). Isolation and purification of a neurotoxin from Bungarus caeruleus (common Indian krait) venom: biochemical changes induced by the toxin in rat. *The Journal of Basic and Clinical Physiology and Pharmacology, 16*, 37–52.

Mukherjee, A. K., Kalita, B., Dutta, S., Patra, A., Maity, C. R., & Punde, D. (2021). Snake envenomation: Therapy and challenges in India. In S. P. Mackessy (Ed.), *Section V: Global approaches to envenomation and treatments, handbook of venoms and toxins of reptiles* (2nd ed.). CRC Press.

Murthy, T. S. N. (1998). The venom system of Indian snakes. In B. D. Sharma (Ed.), *Snakes in India- A Source Book* (pp. 75–81). Asiatic Publishing House.

Oh, A. M. F., Tan, C. H., Ariaranee, G. C., Quraishi, N., & Tan, N. H. (2017). Venomics of Bungarus caeruleus (Indian krait): Comparable venom profiles, variable immunoreactivities among specimens from Sri Lanka, India and Pakistan. *Journal of Proteomics, 164*, 1–18.

Prasarnpun, S., Walsh, J., & Harris, J. (2004). β-Bungarotoxin-induced depletion of synaptic vesicles at the mammalian neuromuscular junction. *Neuropharmacology, 47*, 304–314.

Patra, A., Chanda, A., & Mukherjee, A. K. (2019). Quantitative proteomic analysis of venom from Southern India common krait (Bungarus caeruleus) and identification of poorly immunogenic toxins by immune-profiling against commercial antivenom. *Expert Review of Proteomics, 16*(5), 457–469.

Punde, D. P. (2005). Management of snake-bite in rural Maharashtra: a 10-year experience. *The National Medical Journal, 18*(2), 71–75.

Schetinger, M., Rocha, J. B. T., Ahmed, M., Morsch, V. M., & Schetinger, M. R. C. (2009). Snake venom acetylcholinesterase. In *Handbook of venoms and toxins of reptiles* (pp. 207–219). CRC Press.

Sharma, B. D. (1998a). The venomous Indian snakes. In *snakes in India: A Source Book* (pp. 115–124). Asiatic Publishing House.

Sharma, B. D. (1998b). Fauna of Indian snakes. In *snakes in India: A Source Book* (pp. 87–108). Asiatic Publishing House.

Sharma, B. D. (1998c). Identification of snakes. In *Snakes in India: A Source Book* (pp. 87–108). Asiatic Publishing House.

Sharma, S., Karthikeyan, S., Betzel, C., & Singh, T. P. (1999). Isolation, purification, crystallization and preliminary X-ray analysis of beta 1-bungarotoxin from Bungarus caeruleus (Indian common krait). *Acta Crystallographica, Section D: Biological Crystallography, 55*, 1093–1094.

Vijayaraghavan, B. (2008). *Snakebite: A book for India* (pp. 1–93). The Chennai Snake Park Trust.

Indian Russell's Viper (*Daboia russelii*)

6

Abstract

Indian Russell's viper (RV) is an important member of the "Big Four" dangerous snakes in India with a bite that requires immediate medical attention. On average, adult individuals can be up to 4 ft in length, and their body color is typically yellowish to brown with a pattern of dark, round spots with black-and-white edges. Color variations are very common in this species. RV is found all over Asia, and it is dispersed throughout the Indian subcontinent, though it is less abundant in the northeast region of the country. Based on their morphological features and mitochondrial DNA analyses, RVs are classified into two species: (a) *D. russelii*, which inhabits the Indian subcontinent, and (b) *D. siamensis*, which is prevalent in a few regions of Southeast Asia (except the Indian subcontinent), southern China, Indonesia, and Taiwan. RV is oviparous and a mother can typically give birth to 20–40 offspring. RV venom (RVV) is mainly comprised of proteins and polypeptides that together constitute approximately 90% of the total dry weight of RVV. RVV shows fascinating zoogeographic variation in its composition, as revealed by both biochemical and proteomic analyses. Recently, mass spectrometry analyses were conducted to characterize the enzymatic and nonenzymatic toxin isoforms in RVV samples from different geographical regions of India and Pakistan. The results showed that only nine toxins were common among the RVV samples from the Indian subcontinent. Further, neurotoxic phospholipase A_2 isoforms were identified only in RVV from southern India and to some extent from western India, which can explain the neurotoxic symptoms shown by RV-bite patients from these regions. The lethality (LD_{50} value in mice) of RVV is reported to range from 0.7 (i.v.) to 10 mg/kg (i.p.). The clinical symptoms post-RV bite and the RVV toxins that are responsible for inducing different pathophysiological conditions in experimental animals are described in this chapter.

© The Author(s), under exclusive license to Springer Nature Singapore Pte Ltd. 2021
A. K. Mukherjee, *The 'Big Four' Snakes of India*,
https://doi.org/10.1007/978-981-16-2896-2_6

105

Keywords

Anticoagulant · Composition of Indian Russell's viper venom · Clinical features
of Indian Russell's viper envenomation · *Daboia russelii* · Edema-induction ·
Geographical distribution of Indian Russell's viper · Hemotoxicity · Indian
Russell's viper · Pharmacology of Indian Russell's viper venom · Proteomic
analysis of venom · Variation in Indian Russell's viper venom composition

6.1   Taxonomic Classification of Indian Russell's Viper (*Daboia russelii*)

Phylum: Chordata
Group: Vertebrata
Subphylum: Gnathostomata
Class: Reptilia
Subclass: Diapsida
Order: Squamata
Suborder: Ophidia
Infraorder: Xenophidia
Family: Viperidae
Subfamily: Viperinae
Genus: *Daboia*
Species: *russelii*

6.2   Characteristic Features of the Indian Russell's Viper

Indian Russell's viper (RV) (also known as the chain viper) was named in honor of
Dr. Patrick Russell, a Scottish surgeon and environmentalist who spent a part of his
valuable time in India describing several snakes (Kalita et al., 2018a, b). The genus
name *Daboia* is derived from a Hindi word in the Indian language meaning "that lies
hidden" or "the lurker" (Kalita et al., 2018a). RV is known by several names in
different Indian languages (Table 6.1). The Indian RV (Fig. 6.1) is considered as an
important member of the "Big Four" dangerous snakes with a bite that requires
immediate medical attention (Kalita et al., 2018a; Kalita & Mukherjee, 2019;
Mukherjee et al., 2021).

The size of RV varies from medium to large body size; dorsal scales are keeled
whereas the scales in the lowest row are smooth; it has several different distinct
bright chain patterns; and it has a large triangular head (Khaire, 2014;
Vijayaraghavan, 2008). A detailed description of the morphology of RV is presented
by Mallow et al. (2003). The snake's nostrils are present in the middle of a large,
single nasal scale, and their lower edges touch the nasorostral scale (Mallow et al.,
2003). The supranasal scale is crescent shaped and it divides the nasal scale from the
nasorostral scale anteriorly. The rostral scale is extensive and high (Mallow et al.,

Table 6.1 Vernacular names of Indian Russell's viper

Language	Local name
Bengali	চন্দ্রবোড়া (Chandraborha)
Hindi	दबोईया (Daboia)
Gujarati	ખડચિતડ (Khaḍacitaḍa)
Kannada	ಕೊಳಕು ಮಂಡಳ (kolaku mandala)
Malayalam	അണലി, ചേന തണ്ടൻ (Anali, Chena Thandan)
Marathi	घोणस (Ghonas)
Odia	ବୋଡ଼ା (Boda)
Tamil	கண்ணாடி விரியன் (kanadi virian)
Telegu	రక్త పింజరి (Rakta pinjari)

Fig. 6.1 Photograph of Indian Russell's viper (*Daboia russelii*) (photo courtesy Mr. Vivek Sharma, source: Indiansnakes.org)

2003). The number of mid-body dorsal scales ranges from 27 to 33; the ventral scale or gastrostege represents the distended and crosswise or diagonally stretched-out scales, which extends down the base of the body from the neckline to the anal scale. The number of ventral scales varies from 150 to 180. The undivided anal scale is just in front covering the cloacal opening. The tail is short and covered with 41–68 paired subcaudal scales (Mallow et al., 2003).

The two maxillary bones of RV contain 2–6 pairs of fangs (Mallow et al., 2003). On average, the fangs can enlarge up to 16.5 mm (0.65 in.) (Daniels, 2002). Ernst (1982), however, reported that among the Viperinae family of snakes, the Vipera (*Daboia*) *russelii* fang is intermediate in length, measuring 5.74 mm on average and ranging from 1.5 to 13.0 mm. Newborn RVs have fully functional fangs and some replacement fangs develop with age. In adults, a series of five or six replacement fangs on each side is a normal phenomenon (Ernst, 1982).

The bright and regular spots on the back of RVs are important features as they aid in their easy reorganization (Whitaker & Captain, 2004). RV is a heavy and rough-scaled snake, with vertical eye pupils. In general, they have a bright pattern and their body color is typically yellowish to brown; the pattern is made of dark, round spots

Table 6.2 Body proportions of an adult Indian Russell's viper (Ditmars, 1937)

Body parts	Measurement
Total length	4 ft, 1 in. (124 cm)
Tail length	7 in. (18 cm)
Girth	6 in. (15 cm)
Head (width)	2 in. (5 cm)
Head (length)	2 in. (5 cm)

with black-and-white edges (Whitaker & Captain, 2004, reviewed by Kalita et al., 2018a). The underneath of RV is white in the western part of India, partly dotted in the southeastern, and deeply spotted in northeastern India (Kalita et al., 2018a). Variations in color are common in this species, and they can be best recognized by their short, fat body, triangular shaped head, and regular chain-like pattern (Whitaker & Captain, 2004). Although some similarities are seen between RVs and the fat, harmless common sand boas, the latter possesses shorter and blunter tails and asymmetrical body patterns.

The average length of *D. russelii* is about 120 cm (4 ft) and it can attain a maximum length of 166 cm (5.5 ft) (Whitaker, 2006). In 1937, an American herpetologist and a notable filmmaker Raymond Lee Ditmars described the proportions of a fair-sized adult specimen of RV (Table 6.2).

RV is a terrestrial snake and during the summer and rainy seasons it is primarily active at night (nocturnal forager). Nevertheless, alterations in its behavior have been noted during the winter when its activity increases during daytime. Adult RVs usually display slow and sluggish movement, but when agitated or threatened they become furious, raising 2/3 of their body and producing a loud alarm "hissing" sound to keep strangers away. If pushed to the wall and having no way to escape, they would likely strike and bite. Juveniles are very active and energetic (Kalita et al., 2018a).

Rodents, especially murid species, are the favorite prey of RV though they have also been reported to eat land crabs, tiny reptiles, and scorpions, and juveniles prefer feeding on lizards (Mallow et al., 2003). This species of snake typically dwells in open, grassy, or bushy areas but may also occur in forest plantations and the countryside where prey, such as rodents and lizards, are plentiful (Kalita et al., 2018a; Whitaker, 2006).

6.3　Geographical Distribution, Habitat, and Reproduction of Indian Russell's Viper

RV (*D. russelii*) is found across Asia and dispersed throughout the Indian subcontinent (but less abundantly in the northeast zone of the country). RV is found widely in Southeast Asian countries, including Pakistan, Sri Lanka, Myanmar, Taiwan, and southern China (Kalita et al., 2018a; Mukherjee et al., 2016a; Wüster, 1998; Wüster et al., 1992). RV has been found at elevations above 2756 m (9040 ft) (Whitaker &

Captain, 2004). The snake is classified into five subspecies, based on the variances in body coloration and patterns (Warrell, 1989; reviewed by Kalita et al., 2018a):

1. *Daboia russelii russelii* distributed in India, Nepal, Bangladesh, and Pakistan
2. *Daboia russelii pulchella* found in Sri Lanka
3. *Daboia russelii siamensis* distributed in Thailand, Myanmar, and China
4. *Daboia russelii formosensis* found in Taiwan
5. *Daboia russelii limitis* found in Indonesia

Based on the morphological features and mitochondrial DNA analyses (mitochondrial bar coding), RVs have been classified into two species: (a) *D. russelii* inhabiting the Indian subcontinent and (b) *D. siamensis* prevalent to a few regions in Southeast Asia (except the Indian subcontinent), southern China, Indonesia, and Taiwan (Thorpe et al., 2007; reviewed by Kalita et al., 2018a). Conversely, Tsai et al. (1996) proposed the existence of two types of RV based on the presence of any one of the amino acids: either serine (Ser, S) or asparagine (Asn, N) at the N-terminus of the RV venom phospholipase A_2 (PLA$_2$) isoenzymes. Due to rapid evolution and subsequent divergence of snake venom toxins, this may not be a good taxonomic parameter for classifying RV (Kalita et al., 2018a).

RV is plentiful in eastern, western, and southern regions of India but it is rarely found in the Ganges valley, northern Bengal, or Assam and other states of NE India (Kalita et al., 2018a; Whitaker & Captain, 2004). Generally, RVs are not restricted from dwelling in any particular type of habitation; however, they tend to avoid intense forests. The favorite habitat of RV is in the neighborhood of farmlands and rice cultivation agricultural lands where it can feed on an abundance of its favorite prey, such as rats and mice (Kalita et al., 2018a). For this reason, rice farmers are especially prone to be bitten by RV (Kalita et al., 2018a; Mukherjee, 1998; Mukherjee et al., 2000).

D. russelii is ovoviviparous; when internal fertilization occurs, the mother gives birth to offspring who are born alive (Stidworthy, 1974). The breeding season of RV is generally between spring (April) and the start of the monsoon season (July–August). The gestation period of RV is approximately 7–8 months and the young are born anytime between May and November (frequently around June–July) (Mallow et al., 2003). A mother typically gives birth to 20–40 offspring (Mallow et al., 2003) though, in some cases, a small number of (sometimes just one) offspring will be born (Daniels, 2002). The length of newborns is approximately 215–260 mm (8.5–10.2 in.) and they become sexually mature within 2–3 years after birth (Mallow et al., 2003).

6.4 Composition of Indian Russell's Viper Venom

RV venom (RVV) is predominated by proteins and polypeptides that together constitute approximately 90% of the total dry weight of the venom; the remaining components are inorganic compounds and some metal ions (Mukherjee, 1998). The

Table 6.3 Inorganic and organic constituents of *D. russelii* venom sample from the Burdwan district of West Bengal, eastern India

Parameters	Values
Total nitrogen (%)	17.1
Total phosphorus (μg/100 mg venom)	23.5
Inorganic phosphorus (μg/100 mg venom)	10.0
Acid-soluble phosphorus (μg/100 mg venom)	12.2
Chloride (μg/100 mg venom)	310
Li^+ (μg/100 mg venom)	70
Na^+ (mg/100 mg venom)	2.8
K^+ (μg/100 mg venom)	80
Ca^{2+} (μg/100 mg venom)	175
Total lipid (mg/100 mg venom)	3.0
Total carbohydrate (μg/100 mg venom)	2.3
Total protein (%)	98
(a) Albumin (% of total protein)	60.5
(b) Globulin (% of total protein)	39.5
Free amino acids	Nil
Ribose content (mg/100 mg venom)	1.4

Values are the mean of triplicate determinations (Mukherjee, 1998)

characteristic yellow color of RVV is due to an FAD-containing enzyme L-amino acid oxidase (LAAO) (Mukherjee, 1998; Mukherjee et al., 2015), and the variations in intensity (brightness) of the yellow color may depend on the relative distribution of the enzyme in RVV from different geographical locations. The pH and specific gravity of fresh aqueous solutions of RVV are acidic (~5.8) and range between 1.03 and 1.07, respectively (Devi, 1968; Mukherjee, 1998; reviewed by Kalita et al., 2018a). The inorganic and organic constituents of RVV from eastern India are shown in Table 6.3.

Like other species of snakes, geographical variation in the venom composition of RV is a pervasive phenomenon that was first demonstrated by Jayanthi and Gowda (1988) who used cation-exchange fractionation of RVV from southern India (SI), western India (WI), and northern India (NI) followed by assays of some of the biochemical properties. The disparity in the RVV composition across the country was evident from the different numbers of cation-exchange peaks and the sodium dodecyl sulphate-polyacrylamide gel electrophoresis (SDS-PAGE) protein profiles of the RVV samples collected from the different locales of India. Interestingly, 66 kDa, 39 kDa, and 9 kDa molecular mass toxins were present in NI RVV but absent from SI RVV. In addition, samples of RVV from NI and WI were compared to samples from SI and found to contain higher amounts of acidic PLA_2 enzymes, while the samples from SI were predominated by basic PLA_2 (Jayanthi & Gowda, 1988).

Mukherjee et al. (2000) used Sephadex gel-filtration (GF) chromatography followed by enzymatic assays to characterize the biochemical properties of crude RVV and its GF fractions from eastern India (EI). They established the prevalence of high-molecular-mass (>35 kDa) proteins (toxins) in EI RVV and characterized several enzymes, including the protease, phospholipase A_2 (PLA_2) and procoagulant enzymes. The study also correlated the enzyme activity of EI RVV with its clinical manifestations post-envenomation (Mukherjee et al., 2000). Several laboratories across India have also purified and/or biochemically and pharmacologically characterized numerous enzymes and nonenzymatic proteins from RVV and proposed biomedical applications for some of the toxins (Stocker, 1990; Saikia et al., 2011; Mukherjee et al., 2014a; Mukherjee et al., 2014b; Mukherjee et al., 2014c; Mukherjee et al., 2014d; Mukherjee et al., 2015; Mukherjee, 2013; Thakur et al., 2014, 2016; Kalita et al., 2018c, 2021). Although these studies have significantly contributed to the understanding of the basic complexity and geographical variation in RVV composition, they fail to pinpoint the quantitative variations in RVV composition in the Indian subcontinent (Kalita et al., 2018a). Further biochemical analyses could also help to identify the fraction of nonenzymatic components of RVV.

The well-known toxicity of venom samples, especially RVV, has been determined by using a number of different toxin isoforms and measuring their relative quantitative distribution in venoms (Mukherjee et al., 2000; Stocker et al., 1986). Nevertheless, biochemical analyses were not able to address all of the scientific questions (Kalita et al., 2018a; Kalita & Mukherjee, 2019). To overcome the problem, proteomic (tandem mass spectrometry) analysis of RVV from different locales of India, Sri Lanka and Pakistan was undertaken to elucidate the significant variations in major enzymatic and nonenzymatic proteins (toxins) in samples of RVV (Kalita et al., 2018a; Mukherjee et al., 2016a; Pla et al., 2019; Sharma et al., 2015; Tan et al., 2015) (Table 6.4).

Variation in the composition of RVV is evident from the different proteins (toxins) and protein families identified in RV venom from different geographical regions of the Indian subcontinent (Kalita et al., 2018a). Interestingly, a striking disparity in venom composition between captive and wild RV from Pakistan has been observed (Faisal et al., 2018; Mukherjee et al., 2016a), which suggests that apart from geographical and phylogenetic factors, different living conditions (wild or captive), feeding behaviors, and/or prey items may also contribute to the variation in venom composition (Kalita et al., 2018a).

We conducted a detailed analysis of the RVV proteomes from different locales of the Indian subcontinent, based on the presence of homologous distinct peptides in the toxins (Kalita et al., 2018a). Only nine toxins or proteins (2 Snake venom metalloproteinases (SVMPs), 2 snake venom serine proteases (SVMPs), and single isoforms of Nucleotidase (NT), LAAO, nerve growth factor (NGF), Vascular endothelial growth factor (VEGF), and cysteine-rich secretory protein (CRISP)) were found to be common among the RVV samples from different parts of the country. Moreover, the numbers of unique proteins determined in EI RVV (Burdwan district of West Bengal state), EI RVV (Nadia district of West Bengal state), WI RVV, SI RVV, and Pakistan RVV proteomes were 22, 13, 19, 36, and 25, respectively (Fig. 6.2). The different relative abundance of the protein classes and enzymatic and nonenzymatic toxins of RVV from different locations are depicted in Fig. 6.3a, b (Kalita et al., 2018a).

The proteomics analyses have explicitly identified that enzymatic proteins as compared to nonenzymatic proteins are predominant in RVV. Among the different enzymes, PLA_2, SVMP, and SVSP are the most abundant whereas the Kunitz-type serine protease inhibitor (KSPI) and snaclecs (C-type lectin-like proteins) represent

Table 6.4 Mass spectrometry analysis to characterize the enzymatic and nonenzymatic toxin isoforms in RVV samples from different geographical locations

Protein family	Western India[a]	Eastern India		Southern India[d]	Pakistan (wild specimens)[e]	Pakistan (captive specimens)[f]
		Burdwan[b]	Nadia[c]			
Enzymatic proteins						
PLA$_2$	17 (32.5)	21 (22.2)	12 (21.5)	10 (43.3)	11 (63.8)	17 (32.8)
SVMP	5 (24.8)	10 (19.8)	13 (17.7)	4 (4.6)	3 (2.5)	13 (21.8)
SVSP	6 (8.0)	9 (13.9)	15 (14.3)	18 (12.9)	9 (5.5)	9 (3.2)
LAAO	2 (0.3)	1 (1.7)	2 (1.5)	9 (7.5)	3 (0.8)	2 (0.6)
PDE	1 (1.4)	1 (0.7)	1 (0.5)	2 (1.4)	3 (2.5)	2 (0.6)
NT	2 (0.4)	1 (1.0)	1 (0.5)	2 (1.8)	1 (0.1)	2 (0.6)
Hya	NI (0)	1 (0.1)	1 (0.1)	NI (0)	NI (0)	1 (0.2)
PLB	1 (0.1)	NI (0)	1 (0.1)	1 (1.0)	NI (0)	NI (0)
GC	NI (0)	1 (0.1)	1 (0.4)	1 (1.0)	NI (0)	NI (0)
AMT	NI (0)	NI (0)	NI	NI (0)	NI (0)	1 (0.2)
Apase	NI (0)	1 (0.5)	NI	NI (0)	NI (0)	NI (0)
Total	**34**	**46**	**47**	**47**	**30**	**47**
Nonenzymatic proteins						
KSPI	8 (12.5)	6 (20.3)	5 (22.9)	2 (1.7)	10 (16.0)	8 (28.4)
Snaclec	7 (1.8)	13 (11)	12 (12.1)	11 (14.6)	8 (1.3)	11 (6.4)
CRISP	2 (6.8)	3 (3.9)	2 (3.1)	2 (4.9)	3 (1.3)	3 (2.6)
VEGF	2 (1.8)	2 (1.1)	1 (3.5)	2 (3.7)	2 (4.3)	3 (1.5)
NGF	1 (4.8)	2 (1.6)	1 9 (0.7)	1 (1.6)	2 (1.1)	1 (0.4)
Dis	1 (4.9)	1 (2.2)	1 (1.4)	NI (0)	NI (0)	1 (0.4)
UP	NI (0)	NI	NI	NI (0)	1 (0.7)	1 (0.2)
Total	**21**	**27**	**22**	**18**	**28**	**28**
Grand total	**55**	**73**	**69**	**65**	**58**	**75**

The figures in parentheses represent the percent relative abundance of proteins

NI not identified by LC-MS/MS analysis

[a]Kalita et al. (2017)

[b]Kalita et al. (2018b)

[c]Kalita et al. (2018b)

[d]Kalita et al. (2018d)

[e]Faisal et al. (2018)

[f]Mukherjee et al. (2016a)

PLA2 phospholipase A2, *SMVP* snake venom metalloprotease, *SVSP* snake venom serine protease, *LAAO* L-amino acid oxidase, *PDE* phosphodiesterase, *NT* nucletotidase, *Hya* hyaluronidases, *PLB* phospholipase B, *GC* glutaminyl cyclase, *AMT* aminotransferase, *APase* aminopeptidase, *KSPI* Kunitz-type serine protease inhibitor, *Snaclec* C-type lectin-like proteins, *CRISP* cysteine-rich secretory protein, *VEGF* vascular endothelial growth factor, *NGF* nerve growth factor, Dis disintegrin, *UP* uncharacterized protein

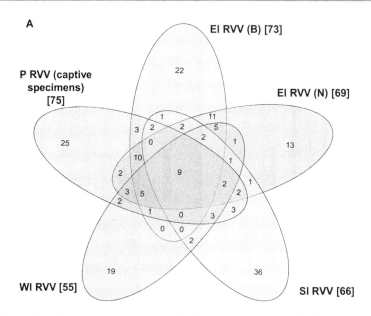

Fig. 6.2 A Venn diagram representing the distribution of common and unique proteins/toxins among RVV samples from different parts of India and Pakistan (Kalita et al., 2018a). Abbreviations: EI RVV (B): eastern India Russel's viper venom (Burdwan district of West Bengal); EI RVV (N): eastern India Russel's viper venom (Nadia district of West Bengal); WI RVV: western India Russel's viper venom; SI RVV: southern India Russel's viper venom; and P RVV: Pakistan Russel's viper venom. The numbers in parentheses indicate the total number of proteins identified in the respective RVV proteomes (this figure and legend are reprinted with permission from Kalita et al., 2018a)

the most copious nonenzymatic toxins of RVV. The proportion of other enzymatic and nonenzymatic toxins is very less in RVV (Table 6.4), and as shown in Table 6.4, the relative proportions of trivial enzymatic classes of RVV, for example, phospholipase B (PLB), carboxypeptidase (CP), aminopeptidase (APase), and glutaminyl cyclase (GC), are comparable among RVV samples from the different regions of the country (Kalita et al., 2018a). The current proteomic analysis is database dependent; therefore, due to paucity of data several enzymatic proteins for example ATPase and ADPase could not be identified by tandem mass spectrometry analysis of RVV but could be detected by biochemical enzyme assay (Mukherjee et al., 2016a; Kalita et al., 2018a, b, d). Further, by biochemical analysis (enzyme assay) a significant variation in the enzyme content in RVV from different regions was observed (Table 6.5).

Various enzymes and nonenzymatic proteins (toxins) purified and characterized from the venom of Indian RV (*D. russelii*) and their biological activity are listed in Tables 6.6 and 6.7, respectively.

Separation of RVV proteins (toxins) by SDS-PAGE under reduced and non-reduced conditions shows their differential migrations, thus suggesting the existence of multiple subunits, noncovalent oligomers (multimeric forms), self-

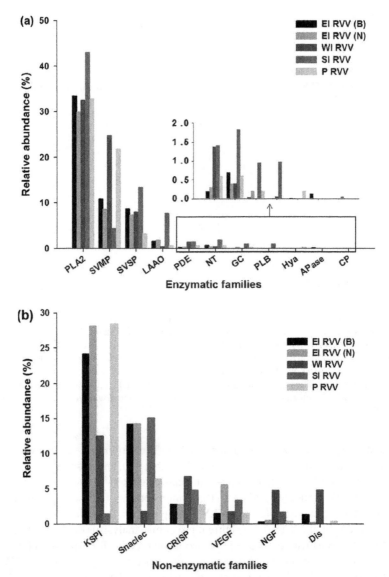

Fig. 6.3 Variation in relative abundance of (**a**) enzymatic proteins and (**b**) nonenzymatic proteins in RVV samples. Abbreviations: EI RVV (B): eastern India Russel's viper venom (Burdwan district of West Bengal); EI RVV (N): eastern India Russel's viper venom (Nadia district of West Bengal); WI RVV: western India Russel's viper venom; SI RVV: southern India Russel's viper venom; and P RVV: Pakistan Russel's viper venom. The numbers in parentheses indicate the total number of proteins identified in the respective RVV proteomes (these figures with legend are reprinted with permission from Kalita et al., 2018a)

Table 6.5 A comparison of enzymatic activities displayed by RVV from different regions of the Indian subcontinent

Enzymatic activity (U/mg)	Origin of RVV samples				
	WT[a]	EI (Burdwan)[b]	EI (Nadia)[c]	SI[d]	Pakistan (captive specimen)[e]
PLA$_2$ (×10^3)[f]	0.6 ± 0.03	0.8 ± 0.02	0.9 ± 0.02	1.1 ± 0.03	0.0078 ± 0.001[g]
SVMP[h]	0.15 ± 0.03	0.10 ± 0.021	0.07 ± 0.011	0.012 ± 0.01	0.2 ± 0.03
LAAO[i]	19.8 ± 0.92	26.7 ± 0.71	24.7 ± 0.6	105.9 ± 2.2	3.2 ± 0.5
Fibrinogenolytic[j]	9.8 ± 0.21	7.6 ± 0.13	5.4 ± 0.11	0.8 ± 0.02	1.5 ± 0.2
Fibrinolytic[j]	0.7 ± 0.04	0.5 ± 0.01	0.3 ± 0.01	0.9 ± 0.01	ND
ATPase (×10^3)[k]	4.5 ± 0.15	1.5 ± 0.05	1.9 ± 0.06	90.0 ± 20.0	4.1 ± 0.8
ADPase (×10^3)[k]	6.4 ± 0.25	2.4 ± 0.09	2.4 ± 0.05	180.0 ± 41.2	ND
AMPase (×10^4)[k]	1.7 ± 0.05	0.5 ± 0.02	0.4 ± 0.02	31.2 ± 8.80	1.2 ± 0.14
Hyaluronidase[l]	63.4 ± 2.11	1918.2 ± 64.1	1946.4 ± 56.3	126.0 ± 2.3	ND
PDE[m]	11.8 ± 0.08	4.5 ± 0.10	4.3 ± 0.12	4.7 ± 0.11	1.2 ± 0.10
TAME (×10^2)[n]	19.1 ± 0.8	3.4 ± 0.11	3.2 ± 0.09	1.6 ± 0.05	ND
BAEE (×10^2)[o]	2.8 ± 0.08	2.0 ± 0.07	2.0 ± 0.06	0.007 ± 0.04	19.1 ± 0.8

Printed with permission from Kalita et al. (2018a)

ND not determined

[a]Kalita et al. (2017)

[b]Kalita et al. (2018b)

[c]Kalita et al. (2018b)

[d]Kalita et al. (2018d)

[e]Mukherjee et al. (2016a)

[f]One unit is defined as a decrease by 0.01 in absorbance at 740 nm after 10 min of incubation

[g]One unit of PLA$_2$ activity of Pakistan RVV is defined as nmol of 3-hydroxy-4-nitrobenzoic acid formed/min/mg protein

[h]One unit is defined as change in absorbance at 450 nm per minute at 37 °C

[i]One unit is defined as 1 nmol of kynurenic acid produced per minute

[j]One unit is defined as 1.0 μg of tyrosine equivalent liberated per minute per mL of enzyme

[k]One unit is defined as micromoles of Pi released per minute

[l]One unit is defined as a 1% decrease in turbidity at 405 nm in comparison to control (100% turbidity)

[m]One unit is defined as micromoles of *p*-nitrophenol released per minute

[n]One unit is defined as an increase by 0.01 in absorbance at 254 nm during the first 5 min of the reaction at 37 °C

[o]One unit is defined as an increase by 0.01 in absorbance at 244 nm during the first 10 min of the reaction at 37 °C

PLA$_2$ Phospholipase A$_2$, *SMVP* snake venom metalloprotease, *LAAO* L-amino acid oxidase, *ATPase* adenosine triphosphatase, *ADPase* adenosine diphosphatase, *AMPase* adenosine monophosphatase, *PDE* phosphodiesterase, *TAME* N-alpha-tosyl-L-arginine methyl ester, *BAEE* Nα-benzoyl-L-arginine ethyl ester

Table 6.6 List of enzymatic proteins purified and characterized from Indian RVV

Enzyme class	Name of enzyme	Molecular weight (kDa)	Biological functions	References
PLA_2	Rv(i) PLA_2	13.6	Anticoagulation, induces inflammation through upregulation of proinflammatory mediators	Deka et al. (2017)
	VRV-PL-V	10	Anticoagulant activity, edema in footpad of mice, neurotoxicity in mice, hemolysis and antibacterial activity against Gram-positive bacteria	Jayanthi and Gowda (1988), Sudarshan and Dhananjaya (2014)
	VRV-PL-V, VRV-PL-VI, and VRV-PL-VIIIa	12–15	LD_{50} is ~5.2 mg/kg body weight of mice, inhibition of ADP-induced platelet aggregation and edema in the footpad of mice. Lacks anticoagulant, myotoxic, and direct hemolytic activities. VRV-PL-VIIIa shows neurotoxicity and myotoxicity, and damages vital organs such as lung, liver, and kidney at its LD_{50} dose	Vishwanath et al. (1988), Kasturi and Gowda (1989), Prasad et al. (1996)
	RV-PL-IIIc, VRV-PL-VII, and VRV-PL-IX	12.5–13.1	VRV-PL-VII and VRV-PL-IX show that LD_{50}s are 7 and 7.5 mg/kg body weight of mice, respectively. VRV-PL-IIIc is not lethal to mice up to 14 mg/kg body weight. They show anticoagulant activity and edema induction in the footpad of mice. VRV-PL-V and VRV-PL-VII are pre- and postsynaptic toxins, while VRV-PL-IX is a presynaptic toxin	Kumar et al. (2015)
	NEUPHOLIPASE	13.0	It shows dose-dependent PLA_2, anticoagulation of platelet-poor mammalian plasma, and indirect hemolysis of mammalian erythrocytes. Demonstrates in vivo liver and erythrocyte damage in experimental mice	Saikia et al. (2013)
	RVVA-PLA_2-I	58.0	Shows strong anticoagulant activity against platelet-poor plasma, and indirect hemolytic activity against mammalian washed erythrocytes, anticoagulation	Saikia et al. (2011, 2012)

		by inhibiting of blood coagulation cascade by both enzymatic and nonenzymatic mechanisms, differential mode of attack on different domains of membrane phospholipids		
	RVV-7	7.2	Cytotoxic activity against B16F10 melanoma cells	Maity et al. (2007)
	RVV-PFIIc'	15.3	LD_{50} (i.p.) of 0.1 mg/kg body weight of mice, anticoagulant activity	Chakraborty et al. (2002)
	RVAPLA$_2$	13.8	Strong anticoagulant by nonenzymatic inhibition of thrombin and plasma phospholipids, no cytotoxicity against mammalian cells in vitro	Mukherjee (2014a)
	Daboxin P	13.6	Shows anticoagulant effect by targeting factor X and Xa	Sharma et al. (2016)
	Viperatoxins VipTx-I VipTx-II	13.6 13.8	VipTx-II demonstrated strong inhibitory (antimicrobial) effect against medically important *Streptococcus aureus* and *Burkholderia pseudomallei* (KHW and TES), *Proteus vulgaris*, and *P. mirabilis* via pore formation and subsequent membrane damage. However, VipTx-I is devoid of the above biological activities	Samy et al. (2015)
	Russelobin	51.2	Thrombin-like protease causes defibrinogenation in vivo resulting in anticoagulation	Mukherjee and Mackessy (2013)
	RV-FVP isoenzymes	32–34 kDa	Activation of factor V, in vivo defibrinogenation resulting in anticoagulation	Mukherjee (2014b)
SVMP	Daborhagin-K	65.06	Induces severe dermal hemorrhage in mice, and may be responsible for showing systemic hemorrhagic symptoms post-RV envenoming	Chen et al. (2008)
	VRR-73	73.0	Fibrinolytic protease with hemorrhagic and esterolytic properties	Chakraborty et al. (2000)
	VRH-1	22.0	Causes severe lung hemorrhage	Chakraborty et al. (1993)

(continued)

Table 6.6 (continued)

Enzyme class	Name of enzyme	Molecular weight (kDa)	Biological functions	References
	RVBCMP	15.0	α-Fibrinogenolytic procoagulant protease; prothrombin activator causes distinct liver hemorrhage	Mukherjee (2008)
	Rusviprotease	26.8	Procoagulant group A-prothrombin activator from snake venom, shows concentration-dependent platelet aggregation and disaggregation, in vivo defibrinogenation in experimental mice	Thakur et al. (2015a)
LAAO	Rusvinoxidase	57	Potent apoptosis induction in MCF-7 cells but comparatively less in Colo-205 cancer cells mainly via intrinsic pathway, nontoxic (4 mg/kg body weight) in experimental mice	Mukherjee et al. (2015, 2018)
	DrLAO	60–64	It inhibits ADP- and collagen-induced platelet aggregation in a dose-dependent manner and the IC_{50} values were determined at 0.27 μM and 0.82 μM, respectively. DrLAO along with other components of RVV acts synergistically to extend and enhance bleeding symptom in RV-bite patients	Chen et al. (2012)
	L1 and L2	60–63	Substrate preference for hydrophobic amino acids; however, no other biological activity is described	Mandal and Bhattacharyya (2008)
PDE	DR-PDE	100	Strong inhibitor of ADP-induced aggregation of platelets	Mitra and Bhattacharyya (2014)
Hyaluronidase	DRHyal-II	28.3	This enzyme is nontoxic; nevertheless it potentiates the myotoxicity of a myotoxin (VRV-PL-VIII) as well as hemorrhagic action of hemorrhagic complex indicating that hyaluronidase thus plays an important role in the pathogenesis by RVV	Mahadeswaraswamy et al. (2011)
ATPase			No biological activity is described	Kini and Gowda (1982)

Apyrase	Ruviapyrase	79.4	Noncytotoxic against breast cancer (MCF-7) cells and non-hemolytic albeit it exhibits marginal anticoagulant and strong antiplatelet activity, dose-dependently reversing the ADP-induced platelet aggregation	Kalita et al. (2018c)

Table 6.7 List of nonenzymatic proteins purified and characterized from Indian RVV

Nonenzymatic protein class	Name of nonenzymatic proteins	Molecular weight (kDa)	Biological functions	References
Kunitz-type serine protease inhibitor (KSPI)	Rusvikunin	6.9	Inhibition of amidolytic activity of trypsin (IC_{50} of 50 nmol/L), plasmin (IC_{50} of 1.1 µmol/L), and fibrinogen clotting as well as plasma-clotting activity of thrombin (IC_{50} of 1.3 µmol/L). Rusvikunin at the tested dose is devoid of lethality in experimental mice, and does not show in vitro cytotoxicity against mammalian cultured cells. It demonstrates in vivo anticoagulant activity in the mouse model	Mukherjee et al. (2014b), Thakur and Mukherjee, (2017a, 2017b)
	Rusvikunin-II	7.1	Shows anticoagulant activity and inhibits the enzymatic (amidolytic) activity of trypsin, plasmin, and FXa in descending order. It also inhibits the fibrinogen-clotting time of thrombin, and, to a smaller extent, the prothrombin activation property of FXa. At a dose of 5.0 mg/kg body weight (i.p. injection) it is found to be nonlethal to mice or house geckos;	Mukherjee and Mackessy (2014), Thakur and Mukherjee (2017a), Kalita and Mukherjee, (2019)

(continued)

Table 6.7 (continued)

Nonenzymatic protein class	Name of nonenzymatic proteins	Molecular weight (kDa)	Biological functions	References
			nevertheless, it shows in vivo anticoagulant action in mice. Rusvikunin-II binds with platelet GPIIb/IIIa receptor via RGD-independent mechanism	
	TI-I	6.8	Synergistic interaction with RVV X to enhance its toxicity and edema-inducing activity in in vivo conditions (mice model)	Jayanthi and Gowda (1990)
	TI-II	7.0	As above	Jayanthi and Gowda (1990)
	DRG-75-U-III	6.5	Inhibits trypsin and component of Reprotoxin complex	Kumar et al. (2008)
C-type lectin (Snaclec)	RVsnaclec	Heterodimer of two subunits, α 15.1 kDa and β 9 kDa	Shows in vitro and in vivo (mice model) anticoagulant activity by inhibition of FXa (Ki = 0.52 μmole) in a Ca^{2+}-independent manner. This toxin does not show hemolytic activity, is devoid of cytotoxicity against human cancer cell lines, demonstrates concentration-dependent aggregation and deaggregation of human platelets, and inhibits the ADP-induced aggregation of	Mukherjee et al. (2014c)

(continued)

Table 6.7 (continued)

Nonenzymatic protein class	Name of nonenzymatic proteins	Molecular weight (kDa)	Biological functions	References
			platelets. RVsnaclec (5.0 mg/kg body weight) is nonlethal to mice and shows no adverse pharmacological effects in experimental mice	
	Daboialectin	18.5	Inhibition of A549 cell growth and morphological alteration, cytoskeletal damage, and apoptosis	Pathan et al. (2017)
Cytotoxin-like proteins	RVV-7	7.2	Induces acute renal failure by both direct and indirect nephrotoxic actions, cytotoxicity	Mandal and Bhattacharyya (2007), Maity et al. (2007)
	Heat-stable protein toxin drCT-I and drCT-II	7.2	Cytotoxicity against mammalian cancer cells	Gomes et al. (2007, 2015)
	Rusvitoxin	6.6	Nontoxic at a dose of 5 mg/kg in mice	Thakur et al. (2015b)
Postsynaptic neurotoxin	DNTx I	6.7	Not determined	Shelke et al. (2002)
Unclassified proteins/ peptides	Pro-angiogenic peptide (RVVAP)	3.9	Noncytotoxic to mammalian cells; however, above 500 ng/ml concentration it can induce chromosomal aberrations, delay in cell cycle kinetics of Chinese hamster ovary cells, angiogenesis, and wound healing determined in experimental animals	Mukherjee et al. (2014d), Thakur et al. (2019)

(continued)

Table 6.7 (continued)

Nonenzymatic protein class	Name of nonenzymatic proteins	Molecular weight (kDa)	Biological functions	References
	Ruviprase	4.4	Potent in vitro and in vivo anticoagulant activity by dual inhibition of thrombin and FXa; shows antiproliferative activity against EMT6/AR1, U-87MG, HeLa, and MCF-7 cancer cells; and is nontoxic to mice at a dose of 2.0 MG/kg	Thakur et al. (2014, 2016)
	VRV-12	12.0	Induction of intense hemorrhage in the intestine but to a lesser extent in the muscles of treated mice	Koley et al. (2000)
Nerve growth factor (NGF)	RVV-NGFa	17.3	Binds with strong affinity to TrkA receptor expressed by PC-12 and breast cancer (MCF-7 and MDA-MB-231) cells. Biological function is unknown	Islam et al. (2020)

aggregation of proteins, and/or interactions among the RVV proteins (Kalita et al., 2017; Mukherjee et al., 2016b) to form protein complexes to augment the pharmacological activity and toxicity of individual RVV proteins (Kalita & Mukherjee, 2019; Mukherjee et al., 2016b; Mukherjee & Mackessy, 2014; Thakur et al., 2015b). A list of protein complexes isolated from RVV and their biological functions is shown in Table 6.8.

6.5 Pharmacology, Pathophysiology, and Clinical Features of Envenomation by Indian Russell's Viper

As described above, RVV shows a fascinating geographical variation in its composition; therefore, its lethality (determined by LD_{50} in mice) is reported to range from 0.7 (i.v.) to 10 mg/kg (i.p.). Studies have shown that among RVV samples from WI,

Table 6.8 A summary of protein (toxin) complexes isolated and characterized from Indian RVV

Toxin components	Name of complex	Molecular weight (kDa)	Biological functions	References
Rusvikunin and Rusvikunin-II (1:2)	Rusvikunin complex	~21 kDa which is self-aggregated to form 60 kDa complex	Anticoagulant activity by thrombin inhibition. Rusvikunin complex at a dose of 5.0 mg/kg is toxic to NSA mice; however, it is nontoxic to house geckos, indicating that it has prey-specific toxicity of this protein complex. The treated mice exhibited dyspnea and hind-limb paresis (muscular weakness due to nerve damage) before death. Therefore, this toxin complex shows neurotoxic symptoms and plays an important role in the assistance of prey subjugation. Rusvikunin complex is bound to and induces RGD-independent aggregation of α-chymotrypsin-treated platelets	Mukherjee and Mackessy (2014), Mukherjee et al. (2016b), Kalita and Mukherjee, (2019)
Phospholipase A_2, protease, and a trypsin inhibitor	Reprotoxin	44.6	It shows lethality in experimental mice with a LD_{50} value of 5.06 mg/kg body weight. This toxin complex inhibits the occurrence of natural release of neurotransmitter in hippocampal neurons. It induces peritoneal bleeding, edema induction in the footpads of mouse, and degeneration of both the germ cells	Kumar et al. (2008)

(continued)

Table 6.8 (continued)

Toxin components	Name of complex	Molecular weight (kDa)	Biological functions	References
			and the Leydig cells of mouse testis, thus showing toxicity to reproductive system	
SVSP and a cytotoxin-like molecule	GF-VI DEAE-II	~35 kDa	It lacks in vitro non-cytotoxicity against mammalian cancer cells, does not induce hemolysis in mammalian erythrocytes, and shows α-fibrin (ogen)ase activity and in vivo toxicity in experimental BALB/c mice (LD_{50} i.p. of 2.3 mg/kg). Within 10 min of i.p. injection this fraction induces fatigue and hind-leg paralysis in mice. It also demonstrates hyperfibrinogenemia, and significantly changes the amounts of factor X, pro- and anti-inflammatory cytokines, viz. TNF-α, IL-6, and IL-10, in the plasma in addition to showing multiple-organ dysfunctions in treated mice	Thakur et al. (2015b)

NI, SI, and EI the latter were the most lethal, suggesting the presence of more lethal or potent toxins in these samples, compared to those from other geographical locations in India (Mukherjee et al., 2000; Jayanthi and Gowda, 1988; Prasad et al., 1999). RV can hold approximately 200–225 mg of venom in its venom glands, so that a large amount of venom can be injected by a single bite into the prey or victim (Kalita et al., 2018a; Saikia et al., 2012). Furthermore, in accordance with the geographical variation in RVV composition, its acute toxicity also varies in the different regions of India. Thus, a disparity is seen in the pharmacological properties and clinical manifestations from RVV in different parts of India, which has been corroborated by proteomic studies on RVV composition (Kalita et al., 2018a; Kalita & Mukherjee, 2019; Mukherjee et al., 2000). For example, SI RVV, compared to RVV samples from P (captive specimens), WI, and EI, shows less procoagulant and fibrin(ogen)olytic activities, which is supported by a reduced cumulative relative abundance of SVMP and SVSP in SI RVV (Kalita et al., 2017; Kalita et al., 2018c; Kalita et al., 2018d; Mukherjee et al., 2016a). Similarly, the relative copiousness of KSPI in SI RVV, compared to that in EI, WI, and P RVV, is meager, which results in its extremely low trypsin inhibitory activity (Jayanthi and Gowda, 1988; Kalita et al., 2018d; Kalita & Mukherjee, 2019). Furthermore, since RVV constituents like snake venom thrombin-like enzyme (SVTLE), snaclec, and LAAO provoke hemostatic disturbances in the prey or victims by platelet aggregation (Rucavado et al., 2001; Stabeli et al., 2004; Xie et al., 2016), proteomic analyses have established that the accumulative relative abundance of these components in SI RVV (23.7%) is greater than the relative distribution of the toxins in RV venoms from WI (2.4%) and EI (13.6–13.8%). Consequently, SI RVV, compared to other RVV samples, induces greater platelet aggregation activity (Kalita et al., 2017; Kalita et al., 2018b; Kalita et al., 2018c; Kalita et al., 2018d).

Proteomic analysis has also shown that RVV is rich in SVMPs (Kalita et al., 2017; Kalita et al., 2018a; Kalita et al., 2018b; Kalita et al., 2018c; Kalita et al., 2018d; Mukherjee et al., 2016a; Sharma et al., 2015; Tan et al., 2015). Typically, this class of enzymes has pharmacological actions such as local ecchymosis, rapid swelling of the bitten part, and extreme pain at the bite site, which can persist for a long time; damage to capillary blood vessels leading to hemorrhage and subsequent capillary leakage syndrome; and formation of intense blisters (outgrowth) over the affected edges (Bawaskar et al., 2008; Gutiérrez et al., 2005; Kalita et al., 2018a; Mukherjee et al., 2000; Mukherjee et al., 2021; Raut & Raut, 2015). Because RVV samples from different localities contain different proportions of SVMPs, the severity of symptoms exhibited by RVV SVMPs in RV-bite patients would also be expected to vary across the country (Kalita & Mukherjee, 2019). The major pharmacological activity of SVSPs and some SVSPs is to affect the vascular system by inducing a hemostatic imbalance, blood coagulation, and hypofibrinogenemia (degradation of plasma fibrinogen) by activating prothrombin, factor X and V, and fibrin (ogen)olytic enzymes that hydrolyze plasma fibrinogen in the victim (Kalita et al., 2018a; Mukherjee, 2008; Mukherjee, 2013; Mukherjee, 2014b; Mukherjee & Mackessy, 2013; Nesheim et al., 1979; Thakur et al., 2015a, 2015b; Thakur & Mukherjee, 2016). These pathophysiological conditions ultimately result in consumptive coagulopathy and incoagulable blood, the most common clinical feature post-RV envenomation (Bawaskar et al., 2008; Kalita et al., 2018a; Mukherjee et al., 2000, 2021; Warrell, 1989). The abundant enzymatic and nonenzymatic anticoagulant toxins of RVV, for example, PLA_2s, KSPIs, and snaclecs, exert their anticoagulating effect by inhibiting thrombin and factor Xa, the key components of blood coagulation that contribute further to incoagulable blood (Mukherjee, 2014a; Mukherjee et al., 2014b, 2014c, 2016a, 2016b). The RGD-dependent binding of RVV KSPI to the platelet GPIIb/IIIa receptor shows an antiplatelet effect in vitro and may also be thrombocytopenia in vivo (Kalita et al., 2019).

The PLA_2 isoenzymes of RVV cause hydrolysis (damage) to the phospholipids of erythrocyte membranes, producing intravascular hemolysis and associated pathological complications in patients bitten by RV (Mukherjee et al., 2000; Saikia et al., 2012; Saikia & Mukherjee, 2017). Therefore, a dramatic variation in quality and relative distribution of PLA_2 isoenzymes in WI, EI, and SI RVV samples is also responsible for the varying extent of indirect hemolysis, intravascular hemolysis, and bleeding problems in RV-envenomed patients from different regions of the country (Kalita et al., 2018a; Kalita & Mukherjee, 2019).

Severe kidney injury, which was previously described as acute renal failure (ARF), is one of the persistent clinical manifestations of RV bite reported in patients across the country. It is likely attributable to LAAO, factor X activators, and PLA_2 isoforms that are present in RVV or the venom of snakes from the Viperidae family (Kalita et al., 2018a; Kalita & Mukherjee, 2019; Morais et al., 2015; Suntravat et al., 2011). Proteomic analyses of RVV have demonstrated significant variations in the relative abundance (quantitative variation) of these components; therefore, differences in the severity of RVV-induced acute kidney injury across the country would be anticipated. Nevertheless, this has not yet been verified due to lack of regional clinical data on snake envenomation (Kalita et al., 2018a). VEGF, another nonenzymatic component of RVV, may worsen the kidney injury and induce strong hypotension in patients. This can lead to a boost in the vascular permeability activities and additional general bleeding problems in RV-bite patients (Kalita et al., 2018a; Yamazaki et al., 2009).

Besides the bleeding complications mentioned above, RV-envenomed patients from SI and occasionally WI have been reported to show unique neuroparalytic symptoms such as ptosis, bulbar palsy, internuclear ophthalmoplegia, and respiratory paralysis due to presynaptic neuromuscular block (Bawaskar et al., 2008; Kalita et al., 2018a; Raut & Raut, 2015; Suchithra et al., 2008). These neurological symptoms are attributed to the neurotoxic PLA_2 isoforms in RVV samples from SI (15.7%) and WI (3.2%) (Kalita et al., 2017; Kalita & Mukherjee, 2019; Kalita et al., 2018c; Kalita et al., 2018d). The correlation between RVV composition and clinical manifestations of RV envenomation in India is shown in Table 6.9.

In conclusion, RV is one of the deadliest venomous snakes of the Indian subcontinent and responsible for a heavy snakebite toll. Like other species of snakes, variation in the venom composition is a common phenomenon that leads to regional differences in the severity of pathogenesis, lethality, and clinical manifestations post-RV envenomation (Kalita et al., 2018a; Kalita & Mukherjee, 2019). Region-specific antivenom would improve the treatment of RV envenomation in different locales of the country. To that end, further studies need to be carried out to understand the clinical manifestations of RV bite in different regions of the country and how the cases are correlated to the composition of RVV in that locality. Specifically, proteomic analyses can fill the knowledge gap on the key components of venom and their effect on the pathophysiology of envenomation.

Table 6.9 Correlation between RVV composition and clinical manifestations of RV envenomation in India

Percent relative abundance in RVV					
		EI[a]			
	Responsible	Burdwan			
Clinical symptoms	toxin(s)	Nadia		WI[b]	SI[c]
Hemostatic imbalance and consumption	SVMP	19.8	17.7	24.8	4.6
coagulopathy	SVSP	13.9	14.3	8.0	12.9
Prolongation in blood coagulation time	PLA$_2$	22.2	21.5	32.5	43.3
	KSPI	20.3	22.9	12.5	1.7
	Snaclec	6.4	12.1	1.8	14.6
Intravascular hemolysis	PLA$_2$	22.2	21.5	32.5	43.3
Edema	SVMP	19.8	17.7	24.8	4.6
	Basic PLA$_2$	4.3	5.2	13.4	41.1
Acute renal failure	RVV-X	9.9	7.1	23.0	3.3
	PLA$_2$	22.2	21.5	32.5	43.3
Neurotoxic symptoms	Neurotoxic PLA$_2$	ND	ND	3.2	19.1

ND not detected
[a]Kalita et al. (2018b)
[b]Kalita et al. (2017)
[c]Kalita et al. (2018c)

References

Bawaskar, H. S., Bawaskar, P. H., Punde, D. P., Inamdar, M. K., Dongare, R. B., & Bhoite, R. R. (2008). Profile of snakebite envenoming in rural Maharashtra, India. *The Journal of the Association of Physicians of India, 56*, 88–95.

Chakraborty, D., Bhattacharyya, D., Sarkar, H. S., & Lahiri, S. C. (1993). Purification and partial characterization of a haemorrhagin (VRH-1) from *Vipera russelli russelli* venom. *Toxicon, 31*, 1601–1614.

Chakraborty, D., Datta, K., Gomes, A., & Bhattacharyya, D. (2000). Haemorrhagic protein of Russell's viper venom with fibrinolytic and esterolytic activities. *Toxicon, 38*, 1475–1490.

Chakraborty, A. K., Hall, R. H., & Ghose, A. C. (2002). Purification and characterization of a potent hemolytic toxin with phospholipase A$_2$ activity from the venom of Indian Russell's viper. *Molecular and Cellular Biochemistry, 237*(1–2), 95–102.

Chanda, A., & Mukherjee, A. K. (2020). Mass spectrometry analysis to unravel the venom proteome composition of Indian snakes: Opening new avenues in clinical research. *Expert Review of Proteomics, 17*, 411–423.

Chen, H.-S., Tsai, H.-Y., Wang, Y.-M., & Tsai, I.-H. (2008). P-III hemorrhagic metalloproteinases from Russell's viper venom: Cloning, characterization, phylogenetic and functional site analyses. *Biochimie, 90*, 1486–1498.

Chen, H. S., Wang, Y. M., Huang, W. T., Huang, K. F., & Tsai, I. H. (2012). Cloning, characterization and mutagenesis of Russell's viper venom L-amino acid oxidase: Insights into its catalytic mechanism. *Biochimie, 94*, 335–344.

Daniels, J. C. (2002). *Book of Indian reptiles and amphibians* (p. 252). Oxford University Press. (Russell's viper, pp. 148–151) ISBN 0-19-566099-4.

Deka, A., Sharma, M., Sharma, M., Mukhopadhyay, R., & Doley, R. (2017). Purification and partial characterization of an anticoagulant PLA$_2$ from the venom of Indian *Daboia russelii* that induces inflammation through upregulation of proinflammatory mediators. *Journal of Biochemical and Molecular Toxicology, 31*(10).

Devi, A. (1968). The protein and non-protein constituents of snake venoms. In *Venomous animals and their venoms* (pp. 119–165). Elsevier.

Ditmars, R. L. (1937) Reptiles of the World: The crocodilians, lizards, snakes, turtles and tortoises of the Eastern and Western Hemispheres, pp: 321, The MacMillan Company, .

Ernst, C. H. (1982). A study of the fangs of Russell's viper (*Vipera russelii*). *Journal of Herpetology, 16*, 67–71.

Faisal, T., Tan, K. Y., Sim, S. M., Quraishi, N., Tan, N. H., & Tan, C. H. (2018). Proteomics, functional characterization and antivenom neutralization of the venom of Pakistani Russell's viper (*Daboia russelii*) from the wild. *Journal of Proteomics, 183*, 1–13.

Gomes, A., Choudhury, S. R., Saha, A., Mishra, R., Giri, B., Biswas, A. K., Debnath, A., & Gomes, A. (2007). A heat stable protein toxin (drCT-I) from the Indian viper (*Daboia russelii russelii*) venom having antiproliferative, cytotoxic and apoptotic activities. *Toxicon, 49*, 46–56.

Gomes, A., Biswas, A. K., Bhowmik, T., Saha, P. P., & Gomes, A. (2015). Russell's Viper venom purified toxin Drct-II inhibits the cell proliferation and induces G1 cell cycle arrest. *Journal of Translational Medicine, 5*. TM (open access).

Gutiérrez, J. M., Rucavado, A., Escalante, T., & Diaz, C. (2005). Hemorrhage induced by snake venom metalloproteinases: Biochemical and biophysical mechanisms involved in microvessel damage. *Toxicon, 45*(8), 997–1011.

Islam, T., Majumdar, M., Bidkar, A., Ghosh, S. S., Mukhopadhyay, R., Utkin, Y., & Mukherjee, A. K. (2020). Nerve growth factor from Indian Russell's viper venom (RVV-NGFa) shows high affinity binding to TrkA receptor expressed in breast cancer cells: Application of fluorescence labeled RVV-NGFa in the clinical diagnosis of breast cancer. *Biochimie, 176*, 311–344.

Jayanthi, G. P., & Gowda, T. V. (1988). Geographical variation in India in the composition and potency of Russell's viper (*Vipera russelli*) venom. *Toxicon, 26*, 257–264.

Jayanthi, G. P., & Gowda, T. V. (1990). Synergistic interaction of a protease and protease inhibitors from Russell's viper (*Vipera russelli*) venom. *Toxicon, 28*(1), 65–74.

Kalita, B., & Mukherjee, A. K. (2019). Recent advances in snake venom proteomics research in India: A new horizon to decipher the geographical variation in venom proteome composition and exploration of candidate drug prototypes. *Journal of Proteins and Proteomics, 10*, 149–164.

Kalita, B., Patra, A., & Mukherjee, A. K. (2017). Unravelling the proteome composition and immuno-profiling of western India Russell's viper venom for in-depth understanding of its pharmacological properties, clinical manifestations, and effective antivenom treatment. *Journal of Proteome Research, 16*, 583–598.

Kalita, B., Mackessy, S. P., & Mukherjee, A. K. (2018a). Proteomic analysis reveals geographic variation in venom composition of Russell's viper in the Indian subcontinent: Implications for clinical manifestations post-envenomation and antivenom treatment. *Expert Review of Proteomics, 15*, 837–849.

Kalita, B., Patra, A., Das, A., & Mukherjee, A. K. (2018b). Proteomic analysis and immuno-profiling of eastern India Russell's viper (*Daboia russelii*) venom: Correlation between RVV composition and clinical manifestations post RV bite. *Journal of Proteome Research, 17*, 2819–2833.

Kalita, B., Patra, A., & Mukherjee, A. K. (2018c). First report of the characterization of a snake venom apyrase (Ruviapyrase) from Indian Russell's viper (*Daboia russelii*) venom. *International Journal of Biological Macromolecules, 111*, 639–648.

Kalita, B., Singh, S., Patra, A., & Mukherjee, A. K. (2018d). Quantitative proteomic analysis and antivenom study revealing that neurotoxic phospholipase A_2 enzymes, the major toxin class of Russell's viper venom from southern India, shows the least immuno-recognition and neutralization by commercial polyvalent antivenom. *International Journal of Biological Macromolecules, 118*, 375–385.

Kalita, B., Dutta, S., & Mukherjee, A. K. (2019). RGD-independent binding of Russell's viper venom Kunitz-type protease inhibitors to platelet GPIIb/IIIa receptor. *Scientific Reports, 9*, 8316.

Kalita, B., Saviola, A. J., & Mukherjee, A. K. (2021). From venom to drugs: A review and critical analysis of Indian snake venom toxins envisaged as anti-cancer drug prototypes. *Drug Discovery Today, 26*(4), 993–1005.

Kasturi, S., & Gowda, T. V. (1989). Purification and characterization of a major phospholipase A_2 from Russell's viper (*Vipera russelii*) venom. *Toxicon, 27*, 229–237.

Khaire, N. (2014). *Indian snakes: A field guide*. Jyotsna Prakasan.

Kini, R. M., & Gowda, T. V. (1982). Studies on snake venom enzymes: Part II—Partial characterization of ATPases from Russell's viper (*Vipera russelii*) venom and their interaction with potassium gymnemate, Indian J. *Biochemical and Biophysical Research Communications, 19*, 342–346.

Koley, L., Chakraborty, D., Datta, K., & Bhattacharyya, D. (2000). Purification and characterization of an organ specific haemorrhagic toxin from *Vipera russelii russelii* (Russell's viper) venom. *Indian Journal of Biochemistry and Biophysics, 37*, 114–120.

Kumar, J. R., Basavarajappa, B. S., Arancio, O., Aranha, I., Gangadhara, N. S., Yajurvedi, H. N., & Gowda, T. V. (2008). Isolation and characterization of "Reprotoxin", a novel protein complex from *Daboia russelii* snake venom. *Biochimie, 90*(10), 1545–1559.

Kumar, J. R., Basavarajappa, B. S., Vishwanath, B. S., & Gowda, T. V. (2015). Biochemical and pharmacological characterization of three toxic phospholipase A_2s from *Daboia russelii* snake venom. *Comparative Biochemistry and Physiology Part C: Toxicology & Pharmacology, 168*, 28–38.

Mahadeswaraswamy, Y. H., Manjula, B., Devaraja, S., Girish, K. S., & Kemparaju, K. (2011). *Daboia russelii* venom hyaluronidase: Purification, characterization and inhibition by β-3-(3-hydroxy-4-oxopyridyl) α-amino-propionic acid. *Current Topics in Medicinal Chemistry, 11*, 2556–2565.

Maity, G., Mandal, S., Chatterjee, A., & Bhattacharyya, D. (2007). Purification and characterization of a low molecular weight multifunctional cytotoxic phospholipase A_2 from Russell's viper venom. *Journal of Chromatography B: Analytical Technologies in the Biomedical and Life Sciences, 845*(2), 232–243.

Mallow, D., Ludwig, D., & Nilson, G. (2003). *True vipers: Natural history and toxinology of old world vipers* (p. 359). Krieger Publishing Company. isbn:0-89464-877-2.

Mandal, S., & Bhattacharyya, D. (2007). Ability of a small, basic protein isolated from Russell's viper venom (*Daboia russelii russelii*) to induce renal tubular necrosis in mice. *Toxicon, 50*, 236–250.

Mandal, S., & Bhattacharyya, D. (2008). Two L-amino acid oxidase isoenzymes from Russell's viper (*Daboia russelii russelii*) venom with different mechanisms of inhibition by substrate analogs. *The FEBS Journal, 275*(9), 2078–2095.

Mitra, J., & Bhattacharyya, D. (2014). Phosphodiesterase from *Daboia russelii russelii* venom: Purification, partial characterization and inhibition of platelet aggregation. *Toxicon, 88*, 1–10.

Morais, I. C. O., Pereira, G. J. S., Orzáez, M., Jorge, R. J. B., Bincoletto, C., Toyama, M. H., Monteiro, H. S. A., Smaili, S. S. S., Pérez-Payá, E., & Martins, A. M. C. (2015). L-amino acid oxidase from *Bothrops leucurus* venom induces nephrotoxicity via apoptosis and necrosis. *PLoS One, 10*(7), e0132569.

Mukherjee, A. K. (1998). *In: Some biochemical properties of cobra and Russell's viper venom and their some biological effects on albino rats*. Ph. D. thesis. Burdwan University, Burdwan.

Mukherjee, A. K. (2008). Characterization of a novel pro-coagulant metalloproteinase (RVBCMP) possessing alpha-fibrinogenase and tissue haemorrhagic activity from venom of *Daboia russelii russelii* (Russell's viper): Evidence of distinct coagulant and haemorrhagic sites in RVBCMP. *Toxicon, 51*(5), 923–933.

Mukherjee, A. K. (2013). An updated inventory on properties, pathophysiology and therapeutic potential of snake venom thrombin-like proteases. In S. Chakraborti & N. S. Dhalla (Eds.),

Proteases in health and disease-advances in biochemistry in health and disease (Vol. 7, pp. 163–180. (ISBN 978-1-4614-9233-7)). Springer.

Mukherjee, A. K. (2014a). A major phospholipase A_2 from *Daboia russelii russelii* venom shows potent anticoagulant action via thrombin inhibition and binding with plasma phospholipids. *Biochimie, 99*, 153–161.

Mukherjee, A. K. (2014b). The pro-coagulant fibrinogenolytic serine protease isoenzymes from *Daboia russelii russelii* venom coagulate the blood through factor V activation: Role of glycosylation on enzymatic activity. *PLoS One, 9*(2), e86823. https://doi.org/10.1371/journal.pone.0086823

Mukherjee, A. K., & Mackessy, S. P. (2013). Biochemical and pharmacological properties of a new thrombin-like serine protease (Russelobin) from the venom of Russell's viper (*Daboia russelii russelii*) and assessment of its therapeutic potential. *Biochimica et Biophysica Acta, 1830*(6), 3476–3488.

Mukherjee, A. K., & Mackessy, S. P. (2014). Pharmacological properties and pathophysiological significance of a Kunitz-type protease inhibitor (Rusvikunin-II) and its protein complex (Rusvikunin complex) purified from *Daboia russelii russelii* venom. *Toxicon, 89*, 55–66.

Mukherjee, A. K., Ghosal, S. K., & Maity, C. R. (2000). Some biochemical properties of Russell's viper (*Daboia russelii*) venom from eastern India: Correlation with clinico-pathological manifestation in Russell's viper bite. *Toxicon, 38*, 163–175.

Mukherjee, A. K., Kalita, B., & Thakur, R. (2014a). Two acidic, anticoagulant PLA_2 isoenzymes purified from the venom of monocled cobra *Naja kaouthia* exhibit different potency to inhibit thrombin and factor Xa via phospholipids independent, non-enzymatic mechanism. *PLoS One, 9*(8), e101334.

Mukherjee, A. K., Mackessy, S. P., & Dutta, S. (2014b). Characterization of a Kunitz-type protease inhibitor peptide (Rusvikunin) purified from *Daboia russelii russelii* venom. *International Journal of Biological Macromolecules, 67*, 154–162.

Mukherjee, A. K., Dutta, S., & Mackessy, S. P. (2014c). A new C-type lectin (RVsnaclec) purified from venom of *Daboia russelii russelii* shows anticoagulant activity via inhibition of FXa and concentration-dependent differential response to platelets in a Ca^{2+}-independent manner. *Thrombosis Research, 134*, 1150–1156.

Mukherjee, A. K., Chatterjee, S., Majumdar, S., Saikia, D., Thakur, R., & Chatterjee, A. (2014d). Characterization of a pro-angiogenic, novel peptide from Russell's viper (*Daboia russelii russelii*) venom. *Toxicon, 77*, 26–31.

Mukherjee, A. K., Saviola, A. J., Burns, P. D., & Mackessy, S. P. (2015). Apoptosis induction in human breast cancer (MCF-7) cells by a novel venom L-amino acid oxidase (Rusvinoxidase) is independent of its enzymatic activity and is accompanied by caspase-7 activation and reactive oxygen species production. *Apoptosis, 20*, 1358–1372.

Mukherjee, A. K., Kalita, B., & Mackessy, S. P. (2016a). A proteomic analysis of Pakistan *Daboia russelii russelii* venom and assessment of potency of Indian polyvalent and monovalent antivenom. *Journal of Proteomics, 144*, 73–86.

Mukherjee, A. K., Dutta, S., Kalita, B., Jha, D. K., Deb, P., & Mackessy, S. P. (2016b). Structural and functional characterization of complex formation between two Kunitz-type serine protease inhibitors from Russell's viper venom. *Biochimie, 128*, 138–147.

Mukherjee, A. K., Saviola, A. J., & Mackessy, S. P. (2018). Cellular mechanism of resistance of human colorectal adenocarcinoma cells against apoptosis-induction by Russell's viper venom L-amino acid oxidase (Rusvinoxidase). *Biochimie, 150*, 8–15.

Mukherjee, A. K., Kalita, B., Dutta, S., Patra, A., Maity, C. R., & Punde, D. (2021). Snake envenomation: Therapy and challenges in India. In S. P. Mackessy (Ed.), *Section V: Global approaches to envenomation and treatments, handbook of venoms and toxins of reptiles* (2nd ed.). CRC Press.

Nesheim, M. E., Taswell, J. B., & Mann, K. G. (1979). The contribution of bovine factor V and factor V_a to the activity of prothrombinase. *Journal of Biological Chemistry, 254*(21), 10952–10962.

Pathan, J., Mondal, S., Sarkar, A., & Chakrabarty, D. (2017). Daboialectin, a C-type lectin from Russell's viper venom induces cytoskeletal damage and apoptosis in human lung cancer cells in vitro. *Toxicon, 127*, 11–21.

Pla, D., Sanz, L., Quesada-Bernat, S., Villalta, M., Baal, J., Chowdhury, M. A. W., León, G., Gutiérrez, J. M., Kuch, U., & Calvete, J. J. (2019). Phylovenomics of Daboia russelii across the Indian subcontinent. Bioactivities and comparative in vivo neutralization and in vitro third-generation antivenomics of antivenoms against venoms from India, Bangladesh and Sri Lanka. *Journal of Proteomics, 15*(207), 103443. https://doi.org/10.1016/j.jprot.2019.103443

Prasad, B. N., Kemparaju, K., Bhatt, K. G., & Gowda, T. V. (1996). A platelet aggregation inhibitor phospholipase A_2 from Russell's viper (*Vipera russelii*) venom: Isolation and characterization. *Toxicon, 34*(10), 1173–1185.

Prasad, B. N., Uma, B., Bhatt, S. K., & Gowda, V. T. (1999). Comparative characterisation of Russell's viper (*Daboia/Vipera russelii*) venoms from different regions of the Indian peninsula. *Biochimica et Biophysica Acta, 1428*(2–3), 121–136.

Raut, S., & Raut, P. (2015). Snake bite management experience in western Maharashtra (India). *Toxicon S, 103*, 89–90.

Rucavado, A., Soto, M., Kamiguti, A. S., Theakston, R. D., Fox, J. W., Escalante, T., & Gutiérrez, J. M. (2001). Characterization of aspercetin, a platelet aggregating component from the venom of the snake *Bothrops asper* which induces thrombocytopenia and potentiates metalloproteinase-induced hemorrhage. *Thrombosis and Haemostasis, 85*(4), 710–715.

Saikia, D., Thakur, R., & Mukherjee, A. K. (2011). An acidic phospholipase A_2 (RVVA-PLA₂-I) purified from *Daboia russelii* venom exerts its anticoagulant activity by enzymatic hydrolysis of plasma phospholipids and by non-enzymatic inhibition of factor Xa in a phospholipids/Ca^{2+} independent manner. *Toxicon, 57*, 841–850.

Saikia, D., Bordoloi, N. K., Chattopadhyay, P., Chocklingam, S., Ghosh, S. S., & Mukherjee, A. K. (2012). Differential mode of attack on membrane phospholipids by an acidic phospholipase A_2 (RVVA-PLA₂-I) from *Daboia russelii* venom. *Bichim Biophys Acta –Biomembrane, 12*, 3149–3157.

Saikia, D., Majumdar, S., & Mukherjee, A. K. (2013). Mechanism of *in vivo* anticoagulant and haemolytic activity by a neutral phospholipase A_2 purified from *Daboia russelii russelii* venom: Correlation with clinical manifestations in Russell's viper envenomed patients. *Toxicon, 76*, 291–300.

Saikia, S., & Mukherjee, A. K. (2017). Anticoagulant and membrane damaging properties of snake venom phospholipase A_2 enzymes. In P. Gopalakrishnakone, H. Inagaki, A. K. Mukherjee, T. R. Rahmy, & C. W. Vogel (Eds.), *Handbook of toxinology, volume – snake venom* (pp. 87–104). Springer Nature.

Samy, P. R., Stiles, B. G., Chinnathambi, A., Zayed, M. E., Alharbi, S. A., Franco, O. L., Rowan, E. G., Kumar, A. P., Lim, L. H. K., & Sethi, G. (2015). Viperatoxin-II: A novel viper venom protein as an effective bactericidal agent. *FEBS Open Bio, 5*, 928–941.

Sharma, M., Das, D., Iyer, J. K., Kini, R. M., & Doley, R. (2015). Unveiling the complexities of *Daboia russelii* venom, a medically important snake of India, by tandem mass spectrometry. *Toxicon, 107*, 266–281.

Sharma, M., Iyer, J. K., Shih, N., Majumder, M., Mattaparthi, V. S., Mukhopadhyay, R., & Doley, R. (2016). Daboxin P, a major phospholipase A_2 enzyme from the Indian *Daboia russelii russelii* venom targets factor X and factor Xa for its anticoagulant activity. *PLoS One, 11*, e0153770.

Shelke, R. R. J., Sathish, S., & Gowda, T. V. (2002). Isolation and characterization of a novel postsynaptic/cytotoxic neurotoxin from *Daboia russelii russelii* venom. *Journal of Peptide Research, 59*, 257–263.

Stabeli, R. G., Marcussi, S., Carlos, G. B., et al. (2004). Platelet aggregation and antibacterial effects of an l-amino acid oxidase purified from *Bothrops alternatus* snake venom. *Bioorganic & Medicinal Chemistry, 12*(11), 2881–2886.

Stidworthy, J. (1974). *Snakes of the World* (Revised ed., p. 160). Grosset & Dunlap, ISBN 0-448-11856-4.

Stocker, K. F. (1990). Composition of snake venoms. In K. F. Stocker (Ed.), *Medical use of snake venom proteins* (pp. 33–56). CRC Press.

Stocker, K. F., Fischer, H., & Brogli, M. (1986). Determination of factor X activator in the venom of the saw-scaled viper (*Echis carinatus*). *Toxicon, 24*(3), 313–315.

Suchithra, N., Pappachan, J. M., & Sujathan, P. (2008). Snakebite envenoming in Kerala, South India: Clinical profile and factors involved in adverse outcomes. *Emergency Medicine Journal, 25*(4), 200–204.

Sudarshan, S., & Dhananjaya, B. L. (2014). Antibacterial potential of a basic phospholipase A2 (VRV-PL-V) of *Daboia russelii pulchella* (Russell's viper) venom. *Biochemistry (Moscow), 79* (11), 1237–1244.

Suntravat, M., Yusuksawad, M., Sereemaspun, A., Perez, J. C., & Nuchprayoon, I. (2011). Effect of purified Russell's viper venom-factor X activator (RVV-X) on renal hemodynamics, renal functions, and coagulopathy in rats. *Toxicon, 58*(3), 230–238.

Tan, N. H., Fung, S. Y., Tan, K. Y., Yap, M. K., Gnanathasan, C. A., & Tan, C. H. (2015). Functional venomics of the Sri Lankan Russell's viper (*Daboia russelii*) and its toxinological correlations. *Journal of Proteomics, 128*, 403–423.

Thakur, R., Kumar, A., Bose, B., Panda, D., Saikia, D., Chattopadhyay, P., & Mukherjee, A. K. (2014). A new peptide (Ruviprase) purified from the venom of *Daboia russelii russelii* shows potent anticoagulant activity via non-enzymatic inhibition of thrombin and factor Xa. *Biochimie, 105*, 149–158.

Thakur, R., Chattopadhyay, D., Ghosh, S. S., & Mukherjee, A. K. (2015a). Elucidation of procoagulant mechanism and pathophysiological significance of a new prothrombin activating metalloprotease purified from *Daboia russelii russelii* venom. *Toxicon, 100*, 1–12.

Thakur, R., Chattopadhyay, D., Ghosh, S. S., & Mukherjee, A. K. (2015b). Biochemical and pharmacological characterization of a toxic fraction and its cytotoxin-like component isolated from Russell's viper (*Daboia russelii russelii*) venom. *Comparative Biochemistry and Physiology, Part C, 168*, 55–65.

Thakur, R., Kini, S., Karkulang, S., Banerjee, A., Chatterjee, P., Chanda, A., Chatterjee, A., Panda, D., & Mukherjee, A. K. (2016). Mechanism of apoptosis induction in human breast cancer MCF-7 cell by Ruviprase, a small peptide from *Daboia russelii russelii* venom. *Chemico-Biological Interactions, 258*, 297–304.

Thakur, R., & Mukherjee, A. K. (2016). Pathophysiological significance and therapeutic implications of Russell's viper venom proteins and peptides affecting blood coagulation. In Y. N. Utkin & A. V. Krivoshein (Eds.), *Snake venoms and envenomation: Modern trends and future prospects* (pp. 93–114). Nova Science Publishers.

Thakur, R., & Mukherjee, A. K. (2017a). A brief appraisal on Russell's viper venom (*Daboia russelii*) proteinases. In P. Gopalakrishnakone, H. Inagaki, A. K. Mukherjee, T. R. Rahmy, & C. W. Vogel (Eds.), *Handbook of toxinology, Volume– snake venom* (pp. 123–144). Springer Nature. https://doi.org/10.1007/978-94-007-6648-8_18-1

Thakur, R., & Mukherjee, A. K. (2017b). Pathophysiological significance and therapeutic applications of snake venom protease inhibitors. *Toxicon, 131*, 37–47.

Thakur, R., Chattopadhyay, P., & Mukherjee, A. K. (2019). The wound healing potential of a pro-angiogenic peptide purified from Indian Russell's viper (*Daboia russelii*) venom. *Toxicon, 165*, 72–82.

Thorpe, R. S., Pook, C. E., & Malhotra, A. (2007). Phylogeography of the Russell's viper (*Daboia russelii*) complex in relation to variation in the colour pattern and symptoms of envenoming. *Herpetological Journal, 17*(4), 209–218.

Tsai, I. H., Lu, P. J., & Su, J. C. (1996). Two types of Russell's viper revealed by variation in phospholipases A_2 from venom of the subspecies. *Toxicon, 34*(1), 99–109.

Vijayaraghavan, B. (2008). *Snakebite: A book for India* (pp. 1–93). The Chennai Snake Park Trust.

Vishwanath, B.S., Kini, R,M,, Gowda, T.V. (1988) Purification and partial biochemical characterization of an edema inducing phospholipase A_2 from *Vipera russelii* (Russell's viper) snake
venom. Toxicon 26, 713–720.

Warrell, D. A. (1989). Snake venoms in science and clinical medicine. 1. Russell's viper: Biology,
venom and treatment of bites. *Transactions of the Royal Society of Tropical Medicine and
Hygiene, 83*(6), 732–740.

Whitaker, R. (2006). *Common Indian snakes: A field guide*. Macmillan Indian Pvt. Ltd..

Whitaker, R., & Captain, A. (2004). *Snakes of India: The field guide* (p. 495). Draco Books. ISBN
81-901873-0-9.

Wüster, W. (1998). The genus *Daboia* (Serpentes: Viperidae): Russell's viper. *Hamadryad-
Madras, 23*, 33–40.

Wüster, W., Otsuka, S., Malhotra, A., & Thorpe, R. S. (1992). Population systematics of Russell's
viper: A multivariate study. *Biological Journal of the Linnean Society, 47*, 97–113.

Xie, H., Huang, M., Hu, Q., Sun, K., Wu, H., Shu, W., Li, X., & Fang, L. (2016). Agkihpin, a novel
SVTLE from *Gloydius halys* Pallas, promotes platelet aggregation *in vitro* and inhibits thrombus
formation *in vivo* in murine models of thrombosis. *Toxicon, 122*, 78–88.

Yamazaki, Y., Matsunaga, Y., Tokunaga, Y., Obayashi, S., Saito, M., & Morita, T. (2009). Snake
venom vascular endothelial growth factors (VEGF-Fs) exclusively vary their structures and
functions among species. *The Journal of Biological Chemistry, 284*(15), 9885–9891.

Indian Saw-Scaled Viper (*Echis carinatus carinatus*)

7

Abstract

The saw-scaled viper (or carpet viper) is endemic to Asia and often found in the Indian subcontinent. Twelve species of *Echis* are found worldwide, though *Echis carinatus carinatus* (the Indian saw-scaled viper) is found in peninsular India and it has occupied a key position in the list of category I medically important snakes in India. *E. carinatus* can grow to be 38–80 cm in length; its color varies from grayish to reddish, olive, or pale brown ground color; and its most common pattern is whitish spots with dark-brown edges. It is distributed in the Indian subcontinent, except in eastern and northeastern regions of the country, Gangetic plains, and Himalayan foothills. On average, the yield of milked venom from an adult saw-scaled viper is approximately 12 mg, but the lethal dose for an adult person is estimated to be only 5 mg. Due to lack of a comprehensive study, the geographical variation in venom composition in *E. carinatus* across Indian states has not been established. The venom proteome composition of *E. c. carinatus* from southern India has been analyzed by tandem mass spectrometry and 90 distinct toxins have been identified (enzymatic and nonenzymatic toxin classes) from 15 snake venom protein families. The relative abundance of different toxins in *E. carinatus* venom is presented in this chapter. The *E. c. carinatus* venom is hemotoxic, but in comparison to Russell's viper venom, *Echis* envenomation has a more pronounced effect on blood coagulation. Acute renal failure, bleeding, and hypotension following *E. carinatus* envenomation are frequently reported toxic effects of the venom.

Keywords

Anticoagulant · Clinical features of Indian saw-scaled viper envenomation · Composition of Indian saw-scaled viper venom · *Echis carinatus carinatus* · Hemotoxicity · Indian saw-scaled viper · Pharmacology of Indian saw-scaled viper venom · Proteomic analysis of snake venom

A. K. Mukherjee, *The 'Big Four' Snakes of India*,
https://doi.org/10.1007/978-981-16-2896-2_7

135

7.1   Taxonomic Classification of the Indian Saw-Scaled Viper (*Echis carinatus carinatus*)

Phylum: Chordata
Group: Vertebrata
Subphylum: Gnathostomata
Class: Reptilia
Subclass: Diapsida
Order: Squamata
Suborder: Ophidia
Infraorder: Xenophidia
Family: Viperidae
Subfamily: Viperinae
Genus: *Echis*
Species: *carinatus*

7.2   Characteristic Features of the Indian Saw-Scaled Viper

Echis, commonly known as the saw-scaled viper (or carpet viper), is endemic to Asia but often found in the arid regions of Africa, the Middle East, and the Indian subcontinent (Sri Lanka, Pakistan, and India) (Mallow et al., 2003). The genus name *Echis* is derived from the Greek word meaning "viper." *Echis* typically shows a threat display by rubbing segments of its body together to produce a "sizzling" warning sound that resembles the "working of a saw" to keep its enemy away. Consequently it was given the name saw-scaled viper. Still, if an enemy or prey approaches within its striking range, the snake will quickly bite its target and deliver a small amount of its potent venom. The genus *Echis* includes some of the species responsible for causing the most snakebite morbidity and mortality in the world, India included (Ali et al., 2004; Patra et al., 2017; Bawaskar & Bawaskar, 2019). Twelve species of *Echis* are found globally; however, *Echis carinatus carinatus* (Fig. 7.1) (i.e., the Indian saw-scaled viper) is found in peninsular India (Casewell et al., 2014). A list of different subspecies of *E. carinatus* that are found in the Indian subcontinent is shown in Table 7.1.

Envenomation by *E. c. carinatus* requires immediate medical treatment, and this species occupies a position in the list of category I medically important snakes in India. It is also a member of the "Big Four" venomous snakes of India. Vernacular names of *Echis* in India are shown in Table 7.2.

The total body length of *E. carinatus* generally ranges from 38 to 60 cm (15 and 31 in.) and rarely exceeds 80 cm in length (Mallow et al., 2003). The head is separated from the neck, and the nose is very short and rounded. The small keeled scales and the enlarged supraocular scale cover the nostrils between the shield and the head. The head is covered with 9–14 interocular and 14–21 circumorbital scales. The eye is separated from the 10–12 supralabials by 1–3 rows of scales (Mallow et al., 2003). The full width of the belly is covered with 143–189 rounded ventral

Fig. 7.1 Photograph of the Indian saw-scaled viper (*Echis carinatus*) (PC: Romulus Whitaker. Reprinted with permission from Patra et al., 2017)

Table 7.1 The subspecies of *E. carinatus* distributed in the Indian subcontinent (Mallow et al., 2003; Kochar et al., 2007)

Subspecies	Distribution in the Indian subcontinent
E. c. astolae	Pakistan (Astola Island)
E. c. carinatus	Peninsular India
E. c. multisquamatus	East to western Pakistan
E. c. sinhaleyus	Sri Lanka
E. c. sochureki	Pakistan, northern India, Rajasthan, and Bangladesh

scales, though the extreme posterior portion contains a single scale (Mallow et al., 2003). The coloration of *Echis* varies from grayish to reddish, olive, or a pale brown ground color, but the most common patterns are whitish spots, with dark brown edges separated by lighter patches (Mallow et al., 2003; Khaire, 2014). A whitish cruciform or trident pattern occurs at the top of the head and a faint stripe covers the eye to the angle of the jaw. The color of belly is whitish to pinkish, and it may have faint or distinct brown dots (Mallow et al., 2003; Khaire, 2014).

7.3 Geographic Distribution, Habitat, Behavior, and Reproduction of the Indian Saw-Scaled Viper

Indian saw-scaled vipers are diversely spread across forests; deserts; semideserts; rain forests; scrub forests; mixed, dry, and moist deciduous forests; and grasslands (Khaire, 2014; Internet source: http://indiansnakes.org/). While this snake is distributed in the Indian subcontinent, it is rarely seen in the eastern and northeastern regions of the country, the Gangetic plains, or the Himalayan foothills (Khaire, 2014). The preferable habitats of *E. carinatus* are sand, rock, soft soil, and bushes. It tends to hide under loose rocks, in deep holes or under fallen decomposing woods,

Table 7.2 Vernacular names of the Indian saw-scaled viper

Language	Local name
Bengali	ফুরসা বোড়া সাপ (Fursa boda sap), বোড়া সাপ (Bora Sap), খুঁদে চন্দ্রবোড়া, বঙ্করাজ (Banka raj)
Hindi	अफई (Aphai)
Gujarati	ઝેરી પડકૂ (Zeri padkoo, Tarachha)
Kannada	ಕಲ್ಲು ಹಾವು (Kallu haavu)
Malayalam	*അണലി (Anali)*
Marathi	फुरसं (phoorsa)
Odia	ଧୂଳି ନାଗ (Dhuli Naga)
Sindhi	ڪپر (Kuppur), جاني (Janndi)
Tamil	சுருட்டை விரியன் (surutai vireyan)
Telegu	చిన్న పింజరా (Chinna pinjara), తోటి పింజరా (Thoti pinjara)

Source: https://en.wikipedia.org/wiki/Echis_carinatus; Daniels (2002)

and under stones in open fields (Sengupta et al., 1994). This snake is mostly active at dusk and at night (diurnal), though it may also show some activity during daylight (Mallow et al., 2003). The snake exhibits slow locomotion but can achieve faster creeping using a sidewise zigzag motion. This snake shows a unique ability to climb in bushes and shrubs, sometimes reaching a height of 2 m above ground (Mallow et al., 2003). This species displays distinct behaviors and may be extremely aggressive in nature (Daniels, 2002).

Lizards, frogs, rodents, and various arthropods, such as scorpions, centipedes, and large insects, are the favorite foods of *E. carinatus* (Daniels, 2002). The availability of prey is the deciding factor for its diet (Mallow et al., 2003). Like Russell's viper, Indian *E. carinatus* is ovoviviparous and females give birth to live offspring generally from April to August. A litter typically consists of 3–15 offspring that are around 115–152 mm in size (Daniels, 2002).

7.4   Composition of the Indian Saw-Scaled Viper Venom

On average, the yield of milked venom from an adult saw-scaled viper (about 0.8–1.0 ft. in length) is approximately 12 mg (Dass et al., 1998; Patra et al., 2017), whereas the lethal dose for an adult victim is estimated to be only 5 mg (Daniels, 2002). The LD_{50} dose of *E. carinatus* in mice has been determined to be 0.24 mg/kg (Vijayaraghavan, 2008). The venom of *Echis* is predominated by proteins and SDS-PAGE analysis has shown the large proteins to be in the mass range of 55–90 kDa albeit the mass range of the prevailing proteins in *E. c. carinatus* venom has been shown to be 10–20 kDa (Patra et al., 2017). Nevertheless, in the absence of a comprehensive study, the geographical variation in venom composition of *E. carinatus* across India was not established although a recent proteomic analysis has highlighted this difference in venom composition (Bhatia & Basudevan, 2020).

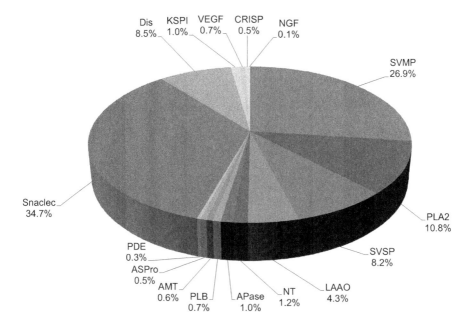

Fig. 7.2 Mass spectrometry analysis of the relative distribution of different toxins in the venom of Indian saw-scaled viper from southern India. The proteomic data was searched against NCBI protein entries with taxonomy set to Viperidae (taxid 8689) (reprinted with permission from Patra et al., 2017). AMT aminotransferase, APase aminopeptidase, ASPro aspartate protease, CRISP cysteine-rich secretory protein, Dis disintegrin, KSPI Kunitz-type serine protease inhibitor, LAAO L-amino acid oxidase, NGF nerve growth factor, NT nucletotidase, PDE phosphodiesterase, PLA2 phospholipase A2, PLB phospholipase B, SMVP snake venom metalloprotease, Snaclec C-type lectin-like proteins, SVSP snake venom serine protease, VEGF vascular

Because the venom yield of this snake is very low, a sufficient quantity is difficult to obtain for research purposes. The venom is also very costly to purchase.

Recently, the venom proteome composition of the Indian saw-scaled viper (*E. c. carinatus*) from southern India was analyzed by tandem mass spectrometry and about 90 distinct toxins (enzymatic and nonenzymatic classes) were identified from 15 snake venom protein families (Patra et al., 2017). For the first time, the proteomic analysis identified the presence of aspartic protease (ASPro), APase, phospholipase B (PLB), vascular endothelial growth factor, and nerve growth factor in Indian saw-scaled viper venom (Patra et al., 2017). The proteome composition of *E. c. carinatus* venom is shown in Fig. 7.2 and a list of proteins (toxins) purified and characterized from Indian saw-scaled viper venom is shown in Table 7.3. Only a few toxins have been purified from this species of snake, and because of the high price and low availability of the venom, few biochemical analyses have been performed.

Another recent study investigated the comparative proteomic analyses of *E. carinatus* venom from different geographical locations of India: Tamil Nadu (ECVTN), Goa (ECVGO), and Rajasthan (ECVRAJ) (Bhatia & Basudevan, 2020). Venom samples were fractionated by Reversed-phase (RP)-HPLC followed by mass spectrometry analysis of the SDS-PAGE protein bands of the RP-HPLC fractions to demonstrate the geographical variation in the venom composition of ECV (Fig. 7.3).

Table 7.3 List of toxins purified and characterized from Indian saw-scaled viper (*E. carinatus*) venom

Enzyme class	Name of enzyme	Molecular weight (kDa)	Biological functions	References
PLA$_2$	EC-IV-PLA2	14	Induces neurotoxicity and edema in mice, but devoid of direct hemolytic, myotoxic, cytotoxic, and anticoagulant activities. LD$_{50}$ (i.p.) is 5 mg/kg body weight of mice	Kemparaju et al. (1994)
	EC-I-PLA2	16	It is nonlethal to mice, does not show neurotoxicity, myotoxicity, anticoagulant activity, and cytotoxicity but induces slight edema in the footpads of experimental mice. It is an inhibitor of ADP, collagen, and epinephrine-induced human platelet aggregation	Kemparaju et al. (1999)
SVMP	EC-PIII	110	Shows in vitro procoagulant activity but devoid of hemorrhagic and myotoxic activities	Choudhury et al. (2018)

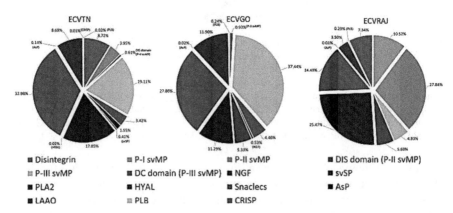

Fig. 7.3 Pie chart representing relative abundance of toxin families for *Echis carinatus* venom: svMP (snake venom metalloproteinases), NGF (neural growth factors), PLA$_2$ (phospholipase A2), PLB (phospholipase B), AsP (renin-like aspartic protease), svSP (snake venom serine proteinase), DIS domain (P-II svMP) (disintegrin domain of P-II svMP), DC domain (disintegrin-like and cysteine domain of P-III svMP), HYAL (hyaluronidase), LAAO (L-amino acid oxidase), CRISP (cysteine-rich secretory protein) (reprinted with permission from Bhatia & Basudevan, 2020). Geographical sources of ECV: Tamil Nadu (ECVTN), Goa (ECVGO), and Rajasthan (ECVRAJ)

The different compositions of ECV between southern India and Sri Lanka have also been demonstrated with only 16 proteins/toxins found to be in common between them (Fig. 7.4) (Patra & Mukherjee, 2020). Due to the geographical variation in the venom composition of ECV, the polyvalent antivenom may be less efficacious in neutralizing the toxicity and pharmacological effects of this venom in different regions of India and Sri Lanka (Patra et al., 2017; Patra & Mukherjee, 2020; Bhatia & Basudevan, 2020).

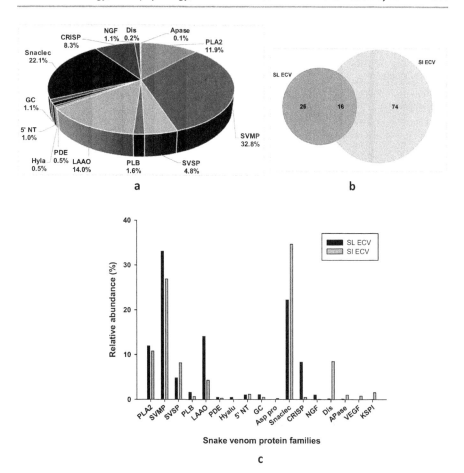

Fig. 7.4 (**a**) Protein family composition of Sri Lanka (SL) ECV proteome. The pie charts represent the relative occurrence of different enzymatic and nonenzymatic protein families of SL ECV when the MS/MS data was searched against the NCBI protein entries with the taxonomy set to Viperidae. (**b**) Venn diagram representing the distribution of common and unique proteins/toxins of SL ECV and southern India (SI) ECV. (**c**) Comparison of relative abundances of protein families of SL ECV and SI ECV (reprinted with permission from Patra & Mukherjee, 2020). APase aminopeptidase, Asp pro aspartate protease, CRISP cysteine-rich secretory protein, Dis disintegrin, GC glutaminyl cyclase, Hya hyaluronidases, KSPI Kunitz-type serine protease inhibitor, LAAO L-amino acid oxidase, NGF nerve growth factor, NT nucletotidase, PDE phosphodiesterase, PLA2 phospholipase A2, PLB phospholipase B, SMVP snake venom metalloprotease, Snaclec C-type lectin-like proteins, SVSP snake venom serine protease, VEGF vascular endothelial growth factor

7.5   Pharmacology, Pathophysiology, and Clinical Features of Envenomation by the Indian Saw-Scaled Viper

An adult saw-scaled viper on average yields approximately 12 mg of venom (Dass et al., 1998). Therefore, approximately 2.5–3.0 µg/mL of venom would be expected to be in the blood of an adult victim after a full bite by *Echis* (Patra et al., 2017).

Table 7.4 Correlation between ECV composition and clinical manifestations of *Echis c. carinatus* envenomation

Clinical symptoms	Responsible toxin (s)	Relative distribution (%)
Hemostatic imbalance and consumption coagulopathy	SVMP	28.2
	SVSP	5.8
Prolongation in blood coagulation time	PLA$_2$	21.7
	KSPI	1.5
	Snaclec	28.6
Intravascular hemolysis	PLA$_2$	21.7
Acute renal failure, edema, and local swelling	SVMP	28.2
	PLA$_2$	21.7
Hemorrhage	SVMP	28.2
Thrombocytopenia and platelet agglutination	Snaclec	28.6

The relative abundance of toxins in ECV (southern India) was determined by proteomic analysis against Viperidae taxid (Patra et al., 2017)

The Viperidae family of snake venoms, by virtue of their effect on the hemostatic system, can cause death in the viper-bite patient (Mukherjee et al., 2000, 2021; Thakur et al., 2015). Nevertheless, compared to Russell's viper venom, *Echis* envenomation produces a more pronounced effect on blood coagulation of the victim (Dass et al., 1998). This has also been demonstrated experimentally (Patra et al., 2017). Local symptoms of *Echis* envenomation are tenacious local swelling, edema, bleeding/hemorrhagic blisters, and pain at the bite site (Kularatne et al., 2011; Patra et al., 2017). Hemotoxicity and coagulopathy, common symptoms of systemic envenomation, have been reported to be severe clinical symptoms in *E. carinatus*-bite patients in different regions of the Indian subcontinent (including southern India), where this snake is prevalent (Dass et al., 1998; Warrell, 1999; Gnanathasan et al., 2012; Patra et al., 2017). Acute renal failure, bleeding, and hypotension following *E. carinatus* envenomation have also been reported (Ali et al., 2004).

Another major problem that is frequently encountered when treating *Echis*-bite patients is the effective management of local swelling. This may not diminish even after administration of several vials of PAV, and victims may become partially crippled (Patra et al., 2017). Another important clinical finding post-*Echis* bite is the significant decrease in circulatory platelets (thrombocytopenia) (Dass et al., 1998; Gnanathasan et al., 2012). Platelet agglutination caused by snaclecs has also been observed (Navdaev et al., 2001).

A correlation between clinical symptoms of *Echis c. carinatus* envenomation and toxins responsible for showing the pathophysiological symptoms and their relative abundance in ECV is shown in Table 7.4.

In conclusion, *Echis c. carinatus* is one of the deadliest snakes in India, with its life-threatening envenomation. Nevertheless, this snake is not prevalent everywhere in the country, and therefore, the antivenom against this snake's venom is not necessarily included in the commercial polyvalent antivenom for treating snakebite in some regions of the country (e.g., eastern and northeastern India), where this snake is not widely found. Another related subspecies (*E. c. sochureki*) is also prevalent in Rajasthan and northern India. Clinical research and especially proteomic analysis are needed to determine the venom composition and potency of commercial antivenoms containing antibodies against *Echis c. carinatus* venom and to neutralize the lethality and toxicity of *E. c. sochureki* venom.

References

Ali, G., Kak, M., Kumar, M., Bali, S. K., Tak, S. I., Hassan, G., & Wadhwa, M. B. (2004). Acute renal failure following *Echis carinatus* (saw-scaled viper) envenomation. *Indian Journal of Nephrology, 14*, 177–181.

Bawaskar, H. S., & Bawaskar, P. H. (2019). Diagnosis of envenomation by Russell's and *Echis carinatus* viper: A clinical study at rural Maharashtra state of India. *Journal of Family Medicine and Primary Care, 8*(4), 1386–1390.

Bhatia, S., & Basudevan, K. (2020). Comparative proteomics of geographically distinct saw-scaled viper (Echis carinatus) venoms from India. *Toxicon: X, 7*, 100048.

Casewell, N. R., Wagstaff, S. C., Wüster, W., Cook, D. A. N., Bolton, F. M. S., King, S. I., Pla, D., Sanz, L., Calvete, J. J., & Harrison, R. A. (2014). Medically important differences in snake venom composition are dictated by distinct postgenomic mechanisms. *Proceedings of the National Academy of Sciences of the United States of America, 111*, 9205–9210.

Choudhury, M., Suvilesh, K. N., Vishwanath, B. S., & Velmurugan, D. (2018). EC-PIII, a novel non-hemorrhagic procoagulant metalloproteinase: Purification and characterization from Indian *Echis carinatus* venom. *International Journal of Biological Macromolecules, 106*, 193–199.

Daniels, J. C. (2002). *Book of Indian reptiles and amphibians* (p. 252). Oxford University Press. Russell's viper, pp. 148–151, ISBN 0-19-566099-4.

Dass, B., Bhatia, R., & Singh, H. (1998). Venomous snakes in India and management of snakebite. In B. D. Sharma (Ed.), *Snakes in India: A Source Book* (pp. 257–268). Asiatic Publishing House.

Gnanathasan, A., Rodrigo, C., Peranantharajah, T., & Coonghe, A. (2012). Saw-scaled viper bites in Sri Lanka: Is it a different subspecies? Clinical evidence from an authenticated case series. *American Journal of Tropical Medicine and Hygiene, 86*, 254–257.

Kemparaju, K., Prasad, B. N., & Gowda, V. T. (1994). Purification of a basic phospholipase A$_2$ from Indian saw-scaled viper (*Echis carinatus*) venom: Characterization of antigenic, catalytic and pharmacological properties. *Toxicon, 32*, 1187–1196.

Kemparaju, K., Krishnakanth, T. P., & Gowda, T. V. (1999). Purification and characterization of a platelet aggregation inhibitor acidic phospholipase A$_2$ from Indian saw-scaled viper (*Echis carinatus*) venom. *Toxicon, 37*, 1659–1671.

Khaire, N. (2014). *Indian snakes: A field guide*. Jyotsna Prakasan.

Kochar, D. K., Tanwar, P. D., Norris, R. L., Sabir, M., Nayak, K. C., Agrawal, T. D., Purohit, V. P., Kochar, A., & Simpson, I. D. (2007). Rediscovery of severe saw-scaled viper (*Echis sochureki*) envenoming in the Thar desert region of Rajasthan, India. *Wilderness & Environmental Medicine, 18*(2), 75–85.

Kularatne, S. A., Sivansuthan, S., Medagedara, S. C., Maduwage, K., & de Silva, A. (2011). Revisiting saw-scaled viper (*Echis carinatus*) bites in the Jaffna Peninsula of Sri Lanka: Distribution, epidemiology and clinical manifestations. *Transactions of the Royal Society of Tropical Medicine and Hygiene, 105*, 591–597.

Mallow, D., Ludwig, D., & Nilson, G. (2003). *True vipers: Natural history and toxinology of old world vipers* (p. 359). Krieger Publishing Company.

Mukherjee, A. K., Ghosal, S. K., & Maity, C. R. (2000). Some biochemical properties of Russell's viper (*Daboia russelii*) venom from eastern India: Correlation with clinico-pathological manifestation in Russell's viper bite. *Toxicon, 38*, 163–175.

Mukherjee, A. K., Kalita, B., Dutta, S., Patra, A., Maity, C. R., & Punde, D. (2021). Snake envenomation: Therapy and challenges in India. In S. P. Mackessy (Ed.), *Section V: Global approaches to envenomation and treatments, handbook of venoms and toxins of reptiles* (2nd ed.). CRC Press.

Navdaev, A., Dörmann, D., Clemetson, J. M., & Clemetson, K. J. (2001). Echicetin, a GPIb-binding snake C-type lectin from *Echis carinatus*, also contains a binding site for IgMκ responsible for platelet agglutination in plasma and inducing signal transduction. *Blood, 97*, 2333–2341.

Patra, A., Kalita, B., Chanda, A., & Mukherjee, A. K. (2017). Proteomics and antivenomics of *Echis carinatus carinatus* venom: Correlation with pharmacological properties and pathophysiology of envenomation. *Scientific Reports, 7*, 17119.

Patra, A., & Mukherjee, A. K. (2020). Proteomic analysis of Sri Lanka *Echis carinatus* venom: Immunological cross-reactivity and enzyme neutralization potency of Indian polyantivenom. *Journal of Proteome Research, 19*, 3022–3032.

Sengupta, S. R., Tare, T. G., Sutar, N. K., & Renapurkar, D. M. (1994). Ecology and distribution of *Echis carinatus* snakes in Devgad Taluka and other areas of Maharashtra State, India. *Journal of Wilderness Medicine, 5*, 282–286.

Thakur, R., Chattopadhyay, D., Ghosh, S. S., & Mukherjee, A. K. (2015). Elucidation of procoagulant mechanism and pathophysiological significance of a new prothrombin activating metalloprotease purified from *Daboia russelii russelii* venom. *Toxicon, 100*, 1–12.

Vijayaraghavan, B. (2008). *Snakebite: A book for India* (pp. 1–93). The Chennai Snake Park Trust.

Warrell, D. A. (1999). The clinical management of snake bites in the Southeast Asian region. *The Southeast Asian Journal of Tropical Medicine and Public Health, S30*, 1–84.

Prevention and Treatment of the "Big Four" Snakebite in India

8

Abstract

Epidemiological survey on rural areas and statistical analyses on hospital data have shown that in India, during the rainy season and throughout the harvesting season, maximum snakebites take place. Therefore, it has been recommended that by adopting some simple precautions and developing some imperative strategies, the human-snake conflicts can easily be avoided. Immediate first aid following a venomous snakebite is extremely essential to save the life of a bite patient followed by which the patient should be transported immediately to a nearby hospital for antivenom treatment. Since maximum snakebites occur in rural areas or villages, the paramedical staff of the tertiary health centers should be trained to deal with providing first aid against snake envenomation. Intravenous administration of monospecific or monovalent and/or polyspecific or polyvalent antivenom (PAV) is the only scientifically accepted therapy against snakebite. However, due to lack of species-specific diagnostic kit, the bitten species of snake cannot be identified accurately; therefore, PAV is widely produced in India and is also supplied to its neighboring countries. The World Health Organization has recommended a guideline for the production and quality assessment of commercial antisnake venom which has been discussed in this chapter. The adverse effects of antivenom including its management and requirement of different doses of antivenom for the treatment of snakebite in India have also been presented in this chapter. Further, species-specific and geographical variation in snake venom composition has a great impact on the effective antivenom treatment and hospital management of snakebite. There is an urgent need for design advances by creating immunization protocols that take into account the above problems and help mitigate the toxic effects of low-molecular-mass, poorly immunogenic snake venom toxic components, leading to the development of region-specific or pan-India antivenom for better in-patient management of snakebite in the Indian subcontinent.

A. K. Mukherjee, *The 'Big Four' Snakes of India*,
https://doi.org/10.1007/978-981-16-2896-2_8

Keywords

Adverse effects of antivenom · Antivenom production · Antivenom therapy · Big Four venomous snakes · Diagnosis of snakebite · First aid for snakebite · Prevention of snakebite · Quality control of antivenom · Serum reactions · Treatment of snakebite

8.1   Prevention of Snakebite: Some Useful Strategies

It has rightly been stated that prevention is better than treatment! Since snakebite is an occupational health hazard mostly affecting the farmers and villagers of rural India, community education might be of great help to make people aware about how to avoid snakes and consequent snakebite. Snakes have a very special habitat and they enter houses mostly in search of food (rat, etc.); therefore, by adopting some simple precautions and developing some important strategies, the human-snake conflicts can easily be avoided. Among the venomous snakes, Indian krait is mostly active during the nighttime (nocturnal or night hunters), whereas Indian cobra and Indian Russell's viper are mainly diurnal and their activity can also be noticed during the daytime. Moreover, epidemiological survey on rural areas and statistical analyses on hospital data have shown that in India, during the rainy season and throughout the harvesting season, maximum snakebites take place (Gaitonde & Bhattacharya, 1980; Gupt et al., 2015; Chandrakumar et al., 2016; Dandong et al., 2018; Mukherjee et al., 2021).

Therefore, it may be suggested that avoiding the habit of sleeping on the floor, particularly during nighttime, use of mosquito nets (that can prevent snakebite as well as mosquito-borne diseases), carrying a flashlight during walking at night or in the dark, wearing gum boots during summer and rainy seasons and while harvesting, not keeping livestock (chicken, ducks) in or near the house, and circumventing spill or open food in the home (that would invite rats and/or mice, the favorite food of snakes) are some of the useful strategies to avoid snakebite. Further, during bicycle riding at night, particularly in the summer and monsoon seasons, one should be careful to avoid trudging over the snakes (moving or lying involuntarily) on the road. The author himself had a slender escape from an unexpected snakebite while cycling intuitively through a dark by-lane of Tezpur town in Assam. Actually, one of the best ways to prevent snakebite is mass community education and creating awareness, by organizing popular lectures and video shows in schools and colleges. The villagers should be trained to avoid snakebite or educated about the measures to be undertaken for handling snakebites.

8.2   First Aid for Snakebite

Undoubtedly, immediate first aid following a venomous snakebite is extremely essential to save the life of a bite patient. First aid provides at least some time to the patient for reaching the nearest hospital for snakebite treatment. If bite victim has received prior training on first aid they can also do it; else it may be accomplished by people attending the sufferer. However, an erroneous misconception on snakebite first aid can be worse and dangerous rather than saving the life of the victim. Some of the examples of such fallacy are making an incision of the bitten part or limb with a knife or blade and then sucking the blood in order to remove the venom (which is shown in many Indian films), rubbing the wound, tying the bitten limb or part with tourniquets, taking the victim to a nearby imposter or traditional healer, and then applying herbs (which are not scientifically proven against snakebite) and magic stones. Of course, the chanting of "mantras" in an effort to cure snakebite is hilarious. Although the effect of *Mantras* is not proven, somewhere it may help to get rid of uneasiness, anxiousness, agony, and pathos in the mind of patients and their family members. Nevertheless, this author in no way supports the use of magic and/or charms which are fake methods of snakebite remedy and urges that affected patients should immediately be provided the recommended first aid (see in subsequent sections) and then be transported immediately to a nearby hospital for antivenom treatment.

8.2.1   First Aid for Snakebite: World Health Organization-Recommended Guidelines

The main aim of first aid is to minimize the spread of venom, prevention of complications of the patient before administrating polyantivenom (PAV) or monoantivenom (MAV), and quick arrangement of transportation of the patient to a nearby tertiary health center or district hospital for medical treatment. Since maximum snakebites occur in rural areas or villages, the paramedical staff of the tertiary health centers should be trained to deal with providing first aid and deal with snake envenomation. Immobilization of bitten limb or bitten part of the patient and lying down him/her in a relaxed position can prevent the contracture of muscle and rapid blood flow to a great extent. It can minimize the absorption of venom into muscles. To be noted, applying tight (arterial) tourniquets is not recommended by the World Health Organization. The most important lifesaving effort of a snakebite victim is moving the concerned patient to a nearby hospital, immediately without wasting a single minute. Post-bite, the first couple of hours is known as "Golden Hour" for snakebite treatment and if antivenin therapy can be started within this time, there is a very high probability that the life of the patient can be saved. However, it is to be remembered that late antivenin therapy may not save the life of a snakebite patient. In fact, counseling of patient is also very effective to reduce the anxiety and deep pain of the patient.

8.3   Antivenom Production in India

Intravenous administration of monospecific or monovalent and/or polyspecific or polyvalent antivenom is the only scientifically accepted therapy against snakebite. In 1880, Albert Calmette at the Institut Pasteur in Saigon introduced equine antisnake serum for the treatment of snake envenomation which was rapidly accepted across the globe. During the recent years antivenom has been categorized under the essential drugs. Several host animals such as hens, rabbit, sheep, goat, camel, and horse may be the target animals for raising antivenom (Landon et al., 1995; Hanly et al., 1995; Landon & Smith, 2003; Lalloo & Theakston, 2003; Cook et al., 2010), but in India horse is the preferred choice for antivenom production because of the following reasons:

1. It is quite easy to handle horses because in general, they are very submissive.
2. Horses can adjust with all the climates of India, and their maintenance is easy.
3. Horses can tolerate administration of repeatedly high doses of snake venoms, which is necessary for raising high titer of IgG in horse plasma.
4. The IgG level in horses is raised rapidly as compared to many other organisms such as rabbit.
5. Horses can produce a large volume of serum.

8.3.1   Monovalent vs. Polyvalent Antivenom

Two types of antivenom are produced commercially: (a) monovalent or monospecific antivenom, and (b) polyvalent or polyspecific. Monovalent antivenom (MAV) is produced against a single species of snake whereas polyvalent antivenom (PAV) in India is produced against a cocktail of venom mixture of "Big Four" (Indian cobra, Indian Russell's viper, Indian common krait, Indian saw-scaled viper) venomous snakes. Therefore, MAV can competently neutralize the venom of that species of snake against which it was raised in horses whereas PAV possessing a cocktail of antibodies against the venoms of "Big Four" snakes of India is capable of neutralizing all of their toxins, however to a different extent. Studies have shown that MAV is better than PAV in showing better immunological cross-reactivity against venoms indicating higher specificity of former antivenom as compared to latter antivenom (Kalita et al., 2017). Undoubtedly, MAV produced against a single species of snake is the most popular choice of treatment for snakebite; however, at this moment a reliable diagnostic kit that can efficiently and specifically diagnose the bitten species of snake is not accessible in India (Mukherjee et al., 2021). Therefore, concerned experts have to rely on their experience and/or textbook knowledge to identify the bitten species of snake before administering the antivenom (Mukherjee et al., 2021). Often, the patient or relatives of the patient may not have seen the snake and hence are oblivious of the morphological feature of the snake that has bitten the patient. Seldom can the kin of the victim bring the snake (live or dead) to hospital for identification. Further, most of the time snakebite patient reaches the hospital after an exhaustive journey and sufficient time and hence may not be available to identify the bitten species of snake (Mohapatra et al., 2011; Mukherjee et al., 2021). Therefore, administration of monovalent antivenom may impose some risk and physicians prefer to administer PAV. As a result, production of MAV in India has significantly declined mainly owing to its commercial non-viability (Mukherjee et al., 2021).

8.3.2 Production of F(ab′)2 PAV in India

India is one of the largest antivenom producer countries of the world. The Indian PAV not only has therapeutic application of mitigating the snakebite problem within the country, but is also exported to its neighboring countries such as Pakistan, Sri Lanka, Bangladesh, Nepal, and Bhutan for snakebite treatment. The current antivenom manufacturing companies of India are listed in Table 8.1.

There are unambiguous guidelines of WHO for the production and quality assessment of commercial antisnake venom (World Health Organization, 2016). Accordingly, equine PAVs manufactured in India contain caprylic acid-precipitated, pepsin-digested F(ab′)2 fragment of immunoglobulin-G (IgG) purified from hyperimmunized plasma of horses (Mukherjee et al., 2021; Patra et al., 2021). The F(ab′)2 fragment is safer than the parent IgG molecule from where it is derived, because the former does not contain the Fc region of IgG and henceforth shows meager serum reactions (León et al., 2013). The F(ab′)2 raised against a specific toxin of venom binds with that toxin to neutralize its toxicity and consequently lethality of the venom. A scheme of PAV production in horses is shown in Fig. 8.1.

Table 8.1 List of antivenom manufacturing companies in India

Manufacturer	Product name	Active molecule
Bharat Serum and Vaccines Limited, Mumbai, Maharashtra, India	Snake Venom Antiserum	Not mentioned
Biological-E Limited, Hyderabad, Telangana, India	Snake Antivenin	F(ab′)2 molecule
Central Research Institute, Kasauli, Himachal Pradesh, India	Polyvalent Antisnake Venom Serum	Not mentioned
Haffkine Bio-Pharmaceutical Corporation Limited, Mumbai, Maharashtra, India	Snake Antivenin	F(ab′)2 molecule
Premium Serums & Vaccines Private Limited, Pune, Maharashtra, India	Snake Venom Antiserum	F(ab′)2 molecule
Virchow Biotech Limited, Hyderabad, Telangana, India	V-ASV	F(ab′)2 molecule

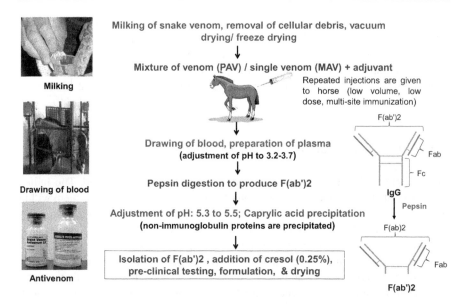

Fig. 8.1 A schematic representation of polyvalent antivenom production in India

8.3.3 Quality Control of Commercial Antivenom: World Health Organization Guidelines

As per the WHO guidelines for the Production, Control and Regulation of Snake Antivenom Immunoglobulins 2016, good manufacturing practice (GMP) should be followed by the antivenom manufacturers from the very beginning to the end of the manufacturing process, as well as for the delivery of the completed product for the maintenance of quality, safety, and clinical efficiency of commercial antivenom. As a matter of fact, particular emphasis has been given on the maintenance of proper record of quality control tests in all stages of production, for the maintenance of transparency in the system (Patra et al., 2021). A proper analysis of the following parameters is suggested by the World Health Organization (2016) before dispensing the bulk product:

1. Assessment of purity and potency of the product
2. Assay of sterility of the product
3. Compliance with the limit of aggregate contents
4. Detection of bacterial endotoxin and pyrogenic contents, if any
5. Concentration of excipients and pH of the formulation

The World Health Organization (2016) has suggested that several standard laboratory tests and preclinical studies should be performed before publicizing the concerned product for quality control of commercial antivenom. The standard quality assays include (a) nonclinical laboratory tests, (b) preclinical tests in experimental rodents (mice/guinea pig), and (c) clinical tests.

8.3.3.1 In Vitro Laboratory Tests

The following simple, but important laboratory tests were recommended by the World Health Organization in 2016 for assuring the quality of commercial antivenom. Further, development of in vitro methods which are well validated to replace the preclinical tests involving animals is highly encouraged by the World Health Organization (2016):

1. Appearance of product
2. Solubility of freeze-dried or lyophilized product
3. Extractable volume of the product from the container
4. Osmolality
5. Identify test by immunological cross-reactivity between antivenom and venom sample(s)
6. Determination of total protein concentration
7. Purity and integrity of immunoglobulin by SDS-PAGE (reduced and non-reduced) or liquid chromatography (size-exclusion FPLC or HPLC) of antivenom
8. Test for pyrogen substances
9. Sterility test to determine that the antivenom formulation is free of bacterial and fungal contaminations
10. Determination of pH and concentrations of sodium chloride, other excipients, and preservatives
11. Determination of residual moisture content of freeze-dried products

8.3.3.2 Preclinical Tests on Experimental Animal Models

The following preclinical tests were recommended by the World Health Organization in 2016. However, it has also been suggested to use minimal number of required animals for the preclinical study:

1. Venom-neutralizing efficacy tests (neutralization of venom lethality) in outbred strains of mice (18–20 g) or in guinea pigs: This should be done for every new antivenom and the new batches of current antivenoms.
2. Abnormal toxicity test at the level of product development.
3. Supplementary preclinical assays, for example, neutralization of hemorrhagic activity, necrotizing activity, procoagulant effect (for Viperidae family of snakes), defibrinogenating activity, myotoxic activity, and neurotoxic activity.

8.3.3.3 Clinical Tests on Volunteers

It has been recommended that clinical tests should be done by following the guidelines set in the international regulations governing good clinical practice: Good Clinical Practice (GCP) Regulations and Guidelines. It is an important prerequisite to register with the appropriate regulatory agency of each country before starting the clinical trials. The conventional pathways for clinical trials of new antivenom are (a) phase I that involves healthy volunteer studies to detect unanticipated adverse effects, (b) phase II that speaks for limited effectiveness and safety studies, (c) phase III that is all about requiring full-scale clinical assessment including blinded randomized clinical trials, and (d) phase IV comprising postmarketing surveillance (World Health Organization, 2016).

8.4 Diagnosis and Clinical Treatment of Snakebite

Due to lack of modern diagnostic kit to determine snake envenomation, the clinicians mostly rely on their knowledge and experience on the determination of snakebite. The identification of snake is in general concluded from the description of the snake, if it can be provided by the patient and/or their family members or eye witness, an examination of the bite site and fang marks, and local symptoms of envenomation, for example, neurotoxic symptoms in case of Elapidae snake bite and hemotoxic symptoms for Viperidae snakebite (Chanda & Mukherjee, 2020; Puzari & Mukherjee, 2020). In addition, biochemical analysis of urine and 20-min whole-blood clotting test (20WBCT) are also followed (Warrell, 1999, 2010). The latter test is a simple and affordable bedside examination for the initial assessment of snakebite and is used to determine the extent of snake envenoming by determining the clinically significant coagulopathy (Sano-Martins et al., 1994; Isbister et al., 2013). In 20WBCT, few milliliters of venous blood are placed in a clean and dry glass test tube and left undisturbed for 20 min and formation of clot indicates positive for snakebite and the test is negative when clot formation does not take place (Isbister et al., 2013). However, this test is applicable only for bite by Viperidae family of snakes showing coagulopathy as the clinical symptom of bite (Isbister et al., 2013; Ratnayake et al., 2017; Raut & Raut, 2015). Nevertheless, 20WBCT has low sensitivity and therefore clinicians cannot rely only on this test to take a decision on antivenom administration albeit classical clinical features of envenomation and results of other bedside tests are also considered by the clinicians before antivenom therapy (Isbister et al., 2013).

Once it is confirmed that the patient is bitten by venomous snakes, intravenous administration of antivenom is the only therapy for hospital management of the snakebite patient (Chanda & Mukherjee, 2020; Mukherjee et al., 2021). However, in their guidelines on antivenom administration, WHO has strongly discouraged the inappropriate use of antivenom, so as not to prevent the development of unnecessary serum reactions in the patient, who does not produce the local and systemic effect of envenomation, but also to stop the antivenom crisis (World Health Organization, 2016). As per the recommendation of WHO, antivenom should be administered to the patient only when he/she shows any one of the following symptoms of envenomation. In case of nonvenomous snakebite or dry bite these symptoms should not appear:

(a) **Local envenomation symptoms:** Swelling of bitten limb (more than half) within 48 h post-bite, rapid extension of swelling, and an enlarged tender lymph node that drains the bitten limb.

(b) **Systemic symptoms of envenoming:** Snake envenomation affecting the hemostatic system of the patient, for example, spontaneous systemic bleeding; positive coagulopathy tests (such as enhanced prothrombin time); neurotoxin symptoms, for example ptosis, paralysis, and difficulty in breathing; cardiovascular dysfunction including hypotension, shock, cardiac arrhythmia (clinical), and irregular electrocardiogram; acute kidney injury or renal failure complications; hemoglobinuria or myoglobinuria characterized by the color of urine that changes to dark brown; intravascular hemolysis; or generalized rhabdomyolysis (Singh et al., 2008; Mukherjee et al., 2021).

Depending on the harshness of envenomation (which depends on several factors) it may require numerous vials of intravenous administration of antivenom until the local sign and systemic effects of envenomation subside (Mukherjee, 2020). Since it has been proved that snakebite is a medical emergency, PAV administration under the supervision of a clinician should be started as early as possible. The number of vials of PAV required for treating a particular snakebite is essentially based upon the awareness of the treating clinician, owing to the fact that snakebite diagnostic kit is unavailable in India and other regions of the world, except Australia (Puzari & Mukherjee, 2020). Therefore, the quantity of snake venom injected into the victim cannot be determined accurately (Chanda & Mukherjee, 2020; Mukherjee et al., 2021). In general, PAV dissolved in physiological saline is administered until local signs of envenomation are diminished. According to physicians, who have the experience of treating a large number of snakebite patients, great strain has been experienced by them to treat the Indian Russell's viper-bite compared to Indian krait-bite and Indian cobra-bite patients. However, as per the physicians, treatment of Indian saw-scaled viper-bite patients is relatively easy (reviewed by Mukherjee et al., 2021).

The treatment dose of PAV for children and pregnant women should be same as that of an adult. Local administration of PAV near the bite site is a futile exercise and does not provide any benefit; instead, patient feels pain and intra-compartmental pressure is increased. For elapid (Indian cobra and Indian krait) envenomation treatment, ten vials of PAV each dissolved in 10 ml of sterile water is administered intravenously within 1 h at an infusion rate of 2 mL/min. However, reappearance of neuroparalysis requires additional 10–15 vials of antivenom management, but even if the neurotoxic symptoms are not reversed, further 5–6 vials of PAV should be administered (Mukherjee et al., 2021). Immediate administration of 0.6 mg atropine and 1 mg neostigmine is recommended if neuro-paresis induced by cobra or krait bite ominously develops. Under emergency, the replication of this treatment is required at an interval of 30 min on an average. If a large amount of cobra venom is injected in patient then approximately 35–40 vials of PAV may be required to reverse the toxic effects of envenomation (Mukherjee et al., 2021). In general, it has been observed that neurotoxic symptoms are improved after the initial 20–30 min of antivenom administration albeit full recovery is achieved after 24–48 h of antivenom administration (Kularatne, 2002; Monteiro et al., 2011; Punde, 2005). Intubation and mechanical ventilation are found to be extremely beneficial in the management of neurotoxic snakebite.

On an average, approximately 4–6 vials of PAV are administered for Indian saw-scaled viper envenomation; however, to reverse the toxicity of Indian Russell's viper (RV) bite ten vials of PAV are initially administered over a period of 1 h and the patient is observed for normalization of blood pressure and pause of spontaneous bleeding within 60 min which are positive indications for effective antivenom therapy (Mukherjee et al., 2021). Nonetheless, the hemostatic disturbance may persist up to several hours. If coagulopathy and spontaneous bleeding continue to occur even after ten vials of PAV administration then additional 8–10 vials of PAV are required to be administered. Administration of up to 40 vials of antivenom may be essential to absolutely alleviate the symptoms post-severe envenomation by Indian RV.

For anonymous vasculotoxic snakebite treatment, initially ten vials of antivenom is administered and the concerned patient is kept under minute observation for additional 1 h (Mukherjee et al., 2021). If the coagulopathy symptoms are not subsidized then additional vials of PAV are administered. Remarkably, effective management of local swelling caused by *Echis* bite is a major challenge for the physicians, because it does not diminish even after several vials of antivenom treatment, thus leaving the victim moderately disabled so the victim is unable to move or walk properly (Patra et al., 2017). Phospholipase A2 (PLA_2) isoenzymes of venom are responsible for edema induction or local swelling and this class of venom enzymes are the poor immunogenic toxins of *Echis* and the other "Big Four" venoms and do not elicit sufficient titer of antibodies in horses (Patra et al., 2017; Kalita et al., 2017; Chanda, Kalita, et al., 2018a). Therefore, commercial PAV is poor in immuno-recognition and neutralization of PLA_2 enzymes (Patra et al., 2017). This may likely explain the reason for the persistence of tenacious local swelling in *Echis*-envenomed patient, even after several vials of antivenom therapy (Patra et al., 2017). Further, venom-induced necrosis or tissue damage and renal failure (kidney damage) are very difficult to treat with antivenom. A recent study by Katkar et al. (2016) suggests that therapeutic administration of DNase I may have additional advantage for preventing the tissue destruction from *Echis* bite. Hemorrhagic snake venom metalloproteases (SVMPs) in Echis carinatus venom (ECV) form a steady neutrophil extracellular trap (NET) at the site of the bite that acts like a barrier to prevent the free flow of blood and thus stops PAV from reaching the damaged site; this is another major challenge for effective treatment of *Echis*-bite patients (Katkar et al., 2016; Patra et al., 2017). Therefore, despite the well immuno-recognition and neutralization of enzyme activity of SVMPs by PAV it fails in preventing and efficient reversing of the ECV-induced local toxicity and swelling (Patra et al., 2017).

PAV treatment post-snakebite may be ineffective to cure ulcers, contractures, and gangrene. Under the circumstances, other than surgical debridement and amputation, practically there are no substitute recourses. The advance treatments are skin grafting and plastic surgery of bitten part, although these are costly affairs and beyond the reach of poor people.

8.5 Management of Adverse Effects of Antivenom

Administration of PAV is associated with adverse reactions, which may be classified into three types: (a) early anaphylactic reactions, (b) endotoxin-mediated pyrogenic reactions, and (c) late serum reactions (De Silva et al., 2015; World Health Organization, 2016).

8.5.1 Early Adverse Reactions

Approximately 20% of the snakebite patients undergoing PAV therapy show early or late serum reactions (Warrell, 1999; World Health Organization, 2016). These are characterized by continuous yawning, appearance of rash, and itching whereas nausea, severe cough with vomiting, urticaria, abdominal colic, diarrhea, tachycardia, tachypnea, bronchospasm, and fever occur within 10 min to 3 h of PAV administration. The early mild anaphylactic reactions occur due to moderate PAV reaction. It has been detected that patients who are highly allergic to equine antibody sometimes develop a severe life-threatening anaphylactic shock (World Health Organization, 2016). The adverse reactions are also dependent on the manufacturing process (De Silva et al., 2015), and therefore, following the good manufacturing practice (GMP) for antivenom production is extremely important, not only to reduce the unwanted serum reactions but also to enhance the potency of commercial antivenom (Patra et al., 2021; World Health Organization, 2016).

8.5.2 Endotoxin-Mediated Pyrogenic Reactions

During the manufacturing process, if the proper precaution is not taken, then the antivenom may be contaminated with Gram-negative bacterial endotoxins responsible for showing pyrogenic reaction in the treated patient. Pyrogenic reaction generally starts after 1–2 h of administration of antivenom treatment and its clinical features are high fever and/or hypertension (Williams, 2007).

8.5.3 Late Serum Reactions

Late serum reaction can be developed with 1–12 days post-antivenom treatment and its symptoms are nausea, vomiting, diarrhea, itching, fever, recurrent urticaria, arthralgia, myalgia, immune complex nephritis, and sometimes encephalopathy (Sharma et al., 2005; Deshpande et al., 2013; Punde, 2005). The adverse clinical reactions of PAV manufactured in India have been studied in those countries which are the frequent users of Indian PAV, for example Sri Lanka, Nepal, and Bangladesh (Theakston et al., 1990; Alirol et al., 2010, 2017; Amin et al., 2008; Stone et al., 2013). However, few fragmentary data are available on the adverse reactions of PAV in India and further clinical studies across the country are necessary to determine the ill effects of antivenom and management of adverse effects of antivenom in India (Deshpande et al., 2013). It has been reported that about 48–89% of PAV-treated patients in different countries (India, Bangladesh, Sri Lanka, and Nepal) showed mild-to-severe acute anaphylactic reactions, for example urticaria (appearance of smooth, red bumps on skin resulting in etching), nausea, vomiting, wheezing, angioedema (swelling of the lower layer of skin or mucous membrane), pyrogenic reaction (fever) with chill, hypotension, bronchospasm without cyanosis, and abrupt respiratory arrest (Theakston et al., 1990; Alirol et al., 2017; Amin et al., 2008; Deshpande et al., 2013). About 5% of patients show late serum reactions (Alirol et al., 2017).

8.5.4   Prevention and Treatment of Adverse Serum Reactions

If a patient undergoing antivenom treatment shows symptoms of early serum reaction then it is better to temporarily suspend the treatment and instantly initiate the adrenaline (epinephrine), antihistaminic, steroids, and/or intravenous (i.v.) fluid therapy (De Silva et al., 2015; Mukherjee et al., 2021). However, in case of severe anaphylaxis reaction, for example bronchospasm and/or abrupt hypotension, the high flow of oxygen and supportive airway and ventilation should be started immediately before the adrenaline administration (De Silva et al., 2015). After diminishing the antiserum reactions, the antivenom infusion should be recommenced however at a slower rate (De Silva et al., 2015). Alternatively, there is no need to wait for the appearance of symptoms of serum reactions and it may be recommended that adrenaline (epinephrine) or antihistaminic drugs may be administered simultaneously with antivenom treatment as a precautionary measurement (Mukherjee et al., 2021). Adrenaline at a dose of 0.01 mg/kg to a maximum of 0.3 mg is administered intramuscularly (i.m.) into the lateral thigh or subcutaneously; if necessary, this low dose can be repeated after 10 min of initial dose (De Silva et al., 2015; Mukherjee et al., 2021). Nevertheless, it has to be kept in mind that Viperidae envenomation causes hemostatic disturbance (coagulopathy) in patients; therefore, extra precaution should be taken for the administration of adrenaline to envenomed patients showing coagulopathy to avoid blood pressure surges; otherwise it may lead to intracerebral hemorrhage (De Silva et al., 2015).

After adrenaline injection, pheniramine maleate (Avil), which is an antihistamine with a little sedative action, is recommended to administer i.m. or i.v. at a dose of 4.5–5.0 mg and 0.35 mg/kg/day for adult and youngster, respectively. Since the administration of antivenom may also cause itching and appearance of rash, an anti-itching drug hydrocortisone (11,17,21-trihydroxypregn-4-ene-3,20-dione) that reduces the swelling, itching, and redness can also be applied at a dose of 100–200 mg (adults) by i.v. injection (De Silva et al., 2015; Mukherjee et al.,

2021). It is extremely important to monitor the blood pressure of the patient at an interval of 3−5 min (De Silva et al., 2015). The nebulized salbutamol and adrenaline are recommended if the patient is showing the symptom of bronchospasm (spasm of bronchial muscle) and upper airway obstruction, respectively, but in case of bradycardia (slow heart rate) i.v. administration of atropine is suggested (De Silva et al., 2015). As soon as the condition of the patient is stabilized, the antivenom therapy should be restarted.

In case of pyrogenic reactions, for example, chills, rigors, fever, myalgia, headache, tachycardia, and hypotension secondary to vasodilation (León et al., 2013), which are mainly caused by bacterial endotoxin and contamination that occur during the first few hours of PAV treatment, paracetamol should be given to the patient. It is recommended that the patient should be monitored for adverse serum reactions, if any, for 72–96 h post-antivenom treatment. If the patient shows late serum reaction, then he/she should be treated with corticosteroids for 1 week, albeit when greater than 25 ml of antivenom is administered to the bite patient it is advisable to give a prophylactic course of oral corticosteroids (Isbister, 2006; De Silva et al., 2015). Notably, patients showing severe late serum sickness should be given oral prednisone (a corticosteroid used to treat swelling and inflammation) at a dose of 60 mg / day and the dose should be reduced gradually over 2 or more weeks to avoid rebound due to sudden withdrawal of steroid (Gold et al., 2002; De Silva et al., 2015).

8.6 Geographical and Species-Specific Variation in Snake Venom Composition and Its Impact on Antivenom Treatment

Geographical and species-specific variation in snake venom composition has a great impact on the pathophysiology post-bite and effective antivenom treatment (Kalita et al., 2018; Mukherjee, 2020; Mukherjee & Maity, 2002; Pla et al., 2019; Senji Laxme et al., 2019; Shashidharamurthy et al., 2010). For example, because of the geographical variation in venom composition equine MAV antivenom produced against the venom of eastern India *N. naja* demonstrated poor effectiveness in neutralizing the lethality and toxicity of western and southern India *N. naja* venoms (Shashidharamurthy et al., 2002; Shashidharamurthy & Kemparaju, 2007). Similarly, PAV manufactured by Haffkine Institute, Maharashtra, was found to be more efficient in neutralizing the toxic effects of western India *N. naja* venom because it was raised against the *N. naja* venom from that region albeit it demonstrated comparatively lower efficacy for neutralizing the toxicity of eastern and southern India *N. naja* venoms. All these findings support that regional differences in cobra venom composition render the PAV partially ineffective (Mukherjee, 2020; Shashidharamurthy & Kemparaju, 2007).

Proteomic analysis has also revealed geographical differences in the venom composition of Indian Russell's viper (RV) across the Indian subcontinent and this variation results in differences in antivenom efficacy against RV envenomation across the country (Kalita et al., 2018; Kalita & Mukherjee, 2019; Pla et al., 2019). This is because almost all the Indian antivenom manufacturers purchase snake venoms from the Irula Snake Catchers Industrial Cooperative Society, Tamil Nadu, southern India (Kalita et al., 2018). However, due to regional variation in snake venom composition antivenom raised against venom of snakes from this particular geographical location might show poor cross-reactivity and neutralization of venom from other regions of the country (Shashidharamurthy & Kemparaju, 2007; Kalita et al., 2018; Mukherjee, 2020).

Species-specific differences in venom composition pose another problem for effective antivenom therapy and hospital management of snakebite. The antivenom raised against a particular species of snake may show poor efficiency in neutralizing the toxicity as well as lethality of another subspecies of snake. For example, severe envenomation due to saw-scaled viper (*Echis sochureki*) in the Thar Desert region of Rajasthan, India, has been reported (Kochar et al., 2007). The difference in venom composition between two subspecies of *E. c. sochureki* and *E. c. carinatus* is unknown; however, PAV was relatively ineffective in restoring coagulation to *E. c. sochureki*-bite patients (Kochar et al., 2007). Likewise, polyvalent antivenom containing antibodies against *N. naja* venom has demonstrated poor efficacy in neutralizing the lethality and toxicity of *N. kaouthia* envenomation in experimental rodents (Mukherjee & Maity, 2002; Mukherjee, 2020). Notably, *N. kaouthia* is the predominant species of cobra in NE India. Recently, Senji Laxme et al. (2019) have also reported the ineffectiveness of PAV in neutralizing the toxicity of medically important regional snake venoms other than the "Big Four" snakes of India. The above findings suggest the requirement of well-designed antivenom by including venom of the same species of snakes from different regions of India. Further, inclusion of antibodies against venom of *N. kaouthia* and/or other medically important local snakes in PAV is also recommended for better treatment against bite by this species of snake (Mukherjee & Maity, 2002; Chanda, Patra, et al., 2018b; Senji Laxme et al., 2019).

In a nutshell, due to the extensive geographical as well as species-specific variation in snake venom composition in India, there is an urgent need for design advances by creating immunization protocols that take into account the above problems and help mitigate the toxic effects of low-molecular-mass, poorly immunogenic snake venom components, leading to the development of region-specific antivenom for better hospital management of snakebite patients (Kalita et al., 2018; Kalita & Mukherjee, 2019). It may further be proposed that a collaborative endeavor among the clinicians, epidemiologists, and basic scientists toward the understanding of snake envenoming and its treatment is vigorous to retaliate this global health issue. Last but not the least, public education for avoidance of snakebite; attention from the government, media, and health authorities; and execution of different schemes for supporting poor patients are the most urgent actions required in today's world (Mukherjee et al., 2021).

References

Alirol, E., Sharma, S. K., Ghimire, A., Poncet, A., Combescure, C., Thapa, C., Paudel, V. P., Adhikary, K., Taylor, W. R., & Warrell, D. (2017). Dose of antivenom for the treatment of snakebite with neurotoxic envenoming: Evidence from a randomised controlled trial in Nepal. *PLOS Neglected Tropical Diseases, 11*, e0005612.

Alirol, E., Sharma, S. K., Bawaskar, H. S., Kuch, U., & Chappuis, F. (2010). Snake bite in South Asia: A review. *PLoS Neglected Tropical Diseases, 4*(1), e603.

Amin, M., Mamun, S., Rashid, R., Rahman, M., Ghose, A., Sharmin, S., Rahman, M., & Faiz, M. (2008). Anti-snake venom: Use and adverse reaction in a snake bite study clinic in Bangladesh. *Journal of Venomous Animals and Toxins including Tropical Diseases, 14*, 660–672.

Chandrakumar, A., Suriyaprakash, T. N. K., LinuMohan, P., Thomas, L., & Vikas, P. V. (2016). Evaluation of demographic and clinical profile of snakebite casualties presented at a tertiary care hospital in Kerala. *Clinical Epidemiology and Global Health, 4*, 140–145.

Chanda, A., Kalita, B., Patra, A., Sandani, W. D., Senevirathne, T., & Mukherjee, A. K. (2018a). Proteomic analysis and antivenomics study of Western India *Naja naja* venom: Correlation between venom composition and clinical manifestations of cobra bite in this region. *Expert Review of Proteomics, 16*(2), 171–184.

Chanda, A., Patra, A., Kalita, B., & Mukherjee, A. K. (2018b). Proteomics analysis to compare the venom composition between *Naja naja* and *Naja kaouthia* from the same geographical location of eastern India: Correlation with pathophysiology of envenomation and immunological cross-reactivity towards commercial polyantivenom. *Expert Review of Proteomics, 15*(11), 949–961.

Chanda, A., & Mukherjee, A. K. (2020). Mass spectrometry analysis to unravel the venom proteome composition of Indian snakes: Opening new avenues in clinical research. *Expert Review of Proteomics, 17*(5), 411–423.

Cook, D. A., Owen, T., Wagstaff, S. C., Kinne, J., Wernery, U., & Harrison, R. A. (2010). Analysis of camelid IgG for antivenom development: Serological responses of venom-immunised camels to separate either monospecific or polyspecific antivenoms for West Africa. *Toxicon, 56*, 363–372.

Dandong, R., Kumar, G. A., Kharyal, A., George, S., Akbar, M., & Dandona, L. (2018). Mortality due to snakebite and other venomous animals in the Indian state of Bihar: Findings from a representative mortality study. *PLoS ONE, 13*(6), e0198900.

De Silva, H. A., Ryan, N. M., & de Silva, H. J. (2015). Adverse reactions to snake antivenom, and their prevention and treatment. *British Journal of Clinical Pharmacology, 81*, 446–452.

Deshpande, R. P., Motghare, V. M., Padwal, S. L., Pore, R. R., Bhamare, C. G., Deshmukh, V. S., & Pise, H. N. (2013). Adverse drug reaction profile of anti-snake venom in a rural tertiary care teaching hospital. *Journal of Young Pharmacists, 5*, 41–45.

Gaitonde, B. B., & Bhattacharya, S. (1980). An epidemiology survey of snakebite cases in India. *The Snake, 12*, 129–133.

Gold, B. S., Dart, R. C., & Barish, R. A. (2002). Bites of venomous snakes. *The New England Journal of Medicine, 347*(5), 347–356.

Gupt, A., Bhatnagar, T., & Murthy, B. N. (2015). Epidemiological profile and management of snakebite cases – A cross sectional study from Himachal Pradesh, India. *Clinical Epidemiology and Global Health, 3*, S96–S100.

Hanly, W. C., Artwohl, J. E., & Bennett, B. T. (1995). Review of polyclonal antibody production procedures in mammals and poultry. *ILAR Journal, 37*(3), 93–118.

Isbister, G. K. (2006). Snake bite: A current approach to management. *Australian Prescriber, 29*, 125–129.

Isbister, G., Maduwage, K., Shahmy, S., Mohamed, F., Abeysinghe, C., Karunathilake, H., Ariaratnam, C., & Buckley, N. (2013). Diagnostic 20-min whole blood clotting test in Russell's viper envenoming delays antivenom administration. *QJM: An International Journal of Medicine, 106*, 925–932.

Kalita, B., & Mukherjee, A. K. (2019). Recent advances in snake venom proteomics research in India: A new horizon to decipher the geographical variation in venom proteome composition and exploration of candidate drug prototypes. *Journal of Proteins and Proteomics, 10*, 149–164.

Kalita, B., Patra, A., & Mukherjee, A. K. (2017). Unravelling the proteome composition and immuno-profiling of western India Russell's viper venom for in-depth understanding of its pharmacological properties, clinical manifestations, and effective antivenom treatment. *Journal of Proteome Research, 16*, 583–598.

Kalita, B., Mackessy, S. P., & Mukherjee, A. K. (2018). Proteomic analysis reveals geographic variation in venom composition of Russell's viper in the Indian subcontinent: Implications for clinical manifestations post-envenomation and antivenom treatment. *Expert Review of Proteomics, 15*, 837–849.

Katkar, G. D., Sundaram, M. S., NaveenKumar, S. K., Swethakumar, B., Sharma, R. D., Paul, M., Vishalakshi, G. J., Devaraja, S., Girish, K. S., & Kemparaju, K. (2016). NETosis and lack of DNase activity are key factors in *Echis carinatus* venom-induced tissue destruction. *Nature Communications, 7*, 11361.

Kochar, D. K., Tanwar, P. D., Norris, R. L., Sabir, M., Nayak, K. C., Agrawal, T. D., Purohit, V. P., Kochar, A., & Simpson, I. D. (2007). Rediscovery of severe saw-scaled viper (*Echis sochureki*) envenoming in the Thar Desert region of Rajasthan, India. *Wilderness & Environmental Medicine, 18*(2), 75–85.

Kularatne, S. (2002). Common krait (*Bungarus caeruleus*) bite in Anuradhapura, Sri Lanka: A prospective clinical study, 1996–98. *Postgraduate Medical Journal, 78*, 276–280.

Lalloo, D. G., & Theakston, R. D. (2003). Snake antivenoms. *Journal of Toxicology. Clinical Toxicology, 41*, 277–290.

Landon, J., Woolley, J. A., & McLean, C. (1995). Antibody production in the hen. In J. Landon & T. Chard (Eds.), *Therapeutic antibodies* (pp. 47–68). Springer-Verlag.

Landon, J., & Smith, D. (2003). Merits of sheep antisera for antivenom manufacture. *Journal of Toxicology: Toxin Reviews, 22*, 15–22.

León, G., Herrera, M., Segura, A., Villalta, M., Vargas, M., & Gutiérrez, J. M. (2013). Pathogenic mechanisms underlying adverse reactions induced by intravenous administration of snake antivenoms. *Toxicon, 76*, 63–76.

Mohapatra, B., Warrell, D. A., Suraweera, W., Bhatia, P., Dhingra, N., Jotkar, R. M., Rodriguez, P. S., Mishra, K., Whitaker, R., & Jha, P. (2011). Million death study, C. snakebite mortality in India: A nationally representative mortality survey. *PLoS Neglected Tropical Diseases, 5*(4), e1018.

Monteiro, F., Kanchan, T., Bhagavath, P., & Kumar, G. P. (2011). Krait bite poisoning in Manipal region of southern India. *Journal of Indian Academy of Forensic Medicine, 33*, 43–45.

Mukherjee, A. K. (2020). Species-specific and geographical variation in venom composition of two major cobras in Indian subcontinent: Impact on polyvalent antivenom therapy. *Toxicon, 188*, 150–158.

Mukherjee, A. K., & Maity, C. R. (2002). Biochemical composition, lethality and pathophysiology of venom from two cobras--*Naja naja* and *N. kaouthia*. *Comparative Biochemistry and Physiology Part B, Biochemistry & Molecular Biology, 131*, 125–132.

Mukherjee, A. K., Kalita, B., Dutta, S., Patra, A., Maity, C. R., & Punde, D. (2021). Snake envenomation: Therapy and challenges in India. In S. P. Mackessy (Ed.), *Section V: Global approaches to envenomation and treatments, handbook of venoms and toxins of reptiles* (2nd ed.). CRC Press.

Patra, A., Kalita, B., Chanda, A., & Mukherjee, A. K. (2017). Proteomics and antivenomics of *Echis carinatus carinatus* venom: Correlation with pharmacological properties and pathophysiology of envenomation. *Nature Scientific Reports, 7*, 17119.

Patra, A., Banerjee, D., Dasgupta, S., & Mukherjee, A. K. (2021). The in vitro laboratory tests and mass spectrometry-assisted quality assessment of commercial polyvalent antivenom raised against the 'Big Four' venomous snakes of India. *Toxicon, 192*, 15–31.

Pla, D., Sanz, L., Quesada-Bernat, S., Villalta, M., Baal, J., Chowdhury, M. A. W., León, G., Gutiérrez, J. M., Kuch, U., & Calvete, J. J. (2019). Phylovenomics of Daboia russelii across the

Indian subcontinent. Bioactivities and comparative in vivo neutralization and in vitro third-generation antivenomics of antivenoms against venoms from India, Bangladesh and Sri Lanka. *Journal of Proteomics, 15*(207), 103443.

Puzari, U., & Mukherjee, A. K. (2020). Recent developments in diagnostic tools and bioanalytical methods for analysis of snake venom: A critical review. *Analytica Chimica Acta, 1137*, 208–224.

Punde, D. P. (2005). Management of snake-bite in rural Maharashtra: A 10-year experience. *National Medical Journal of India, 18*(2), 71–75.

Ratnayake, I., Shihana, F., Dissanayake, D. M., Buckley, N. A., Maduwage, K., & Isbister, G. K. (2017). Performance of the 20-minute whole blood clotting test in detecting venom induced consumption coagulopathy from Russell's viper (Daboia russelii) bites. *Thrombosis and Haemostasis, 117*, 500–507.

Raut, S., & Raut, P. (2015). Snake bite management experience in western Maharashtra (India). *Toxicon, 103*, 89–90.

Sano-Martins, I., Fan, H., Castro, S., Tomy, S., França, F., Jorge, M., Kamiguti, A., Warrell, D., & Theakston, R. (1994). Reliability of the simple 20 minute whole blood clotting test (WBCT20) as an indicator of low plasma fibrinogen concentration in patients envenomed by Bothrops snakes. *Toxicon, 32*, 1045–1050.

Senji Laxme, R. R., Khochare, S., de Souza, H. F., Ahuja, B., Suranse, V., Martin, G., Whitaker, R., & Sunagar, K. (2019). Beyond the 'big four': Venom profiling of the medically important yet neglected Indian snakes reveals disturbing antivenom deficiencies. *PLoS Neglected Tropical Diseases, 13*, e0007899.

Shashidharamurthy, R., Jagadeesha, D. K., Girish, K. S., & Kemparaju, K. (2002). Variations in biochemical and pharmacological properties of Indian cobra (*Naja naja naja*) venom due to geographical distribution. *Molecular and Cellular Biochemistry, 229*(1–2), 93–101.

Shashidharamurthy, R., & Kemparaju, K. (2007). (2007) Region-specific neutralization of Indian cobra (*Naja naja*) venom by polyclonal antibody raised against the eastern regional venom: A comparative study of the venoms from three different geographical distributions. *International Immunopharmacology, 7*(1), 61–69.

Shashidharamurthy, R., Mahadeswaraswamy, Y. H., Ragupathi, L., Vishwanath, B. S., & Kemparaju, K. (2010). Systemic pathological effects induced by cobra (*Naja naja*) venom from geographically distinct origins of Indian peninsula. *Experimental and Toxicologic Pathology, 62*(6), 587–592.

Stone, S. F., Isbister, G. K., Shahmy, S., Mohamed, F., Abeysinghe, C., Karunathilake, H., et al. (2013). Immune response to snake envenoming and treatment with antivenom; complement activation, cytokine production and mast cell degranulation. *PLoS Neglected Tropical Diseases, 7*(7), e2326.

Theakston, R., Phillips, R., , D., Galagedera, Y., Abeysekera, D., Dissanayaka, P., de Silva, A., Aloysius, D. (1990) Envenoming by the common krait (*Bungarus caeruleus*) and Sri Lankan cobra (*Naja naja naja*): Efficacy and complications of therapy with Haffkine antivenom, Transactions of the Royal Society of Tropical Medicine and Hygiene 84, 301–308.

Sharma, N., Chauhan, S., Faruqi, S., Bhat, P., & Varma, S. (2005). Snake envenomation in a north Indian hospital. *Emergency Medicine Journal, 22*(2), 118–120.

Singh, J., Bhoi, S., Gupta, V., & Goel, A. (2008). Clinical profile of venomous snake bites in north Indian military hospital. *Journal of Emergencies, Trauma, and Shock, 1*(2), 78–80.

Warrell, D. A. (1999). The clinical management of snake bites in the Southeast Asian region. *Southeast Asian Journal of Tropical Medicine and Public Health, S30*, 1–84.

Warrell, D. A. (2010). Snake bite. *The Lancet, 375*(9708), 77–88.

Williams, K. L. (2007). *Endotoxins: pyrogens, LAL testing and depyrogenation* (3rd ed.). Informa Healthcare.

World Health Organization. (2016). WHO Guidelines for the Production, Control and Regulation of Snake Antivenom Immunoglobulins. https://www.who.int/bloodproducts/snake_antivenoms/snakeantivenomguide/en/.

Printed by Books on Demand, Germany